# Quality and Process Improvement

Mark A. Fryman

**DELMAR**

™

**THOMSON LEARNING**

Australia   Canada   Mexico   Singapore   Spain   United Kingdom   United States

In dedication to God, who gave me life, and
to my Mother, who totally inspired my life

Quality and Process Improvement
by
Mark A. Fryman

**Business Unit Director:**
Alar Elken
**Executive Editor:**
Sandy Clark
**Acquisitions Editor:**
Mark Huth
**Development:**
Dawn Daugherty

**Executive Production Manager:**
Mary Ellen Black
**Production Editor:**
Ruth Fisher

**Executive Marketing Manager:**
Maura Theriault
**Marketing Coordinator:**
Brian McGrath

COPYRIGHT © 2002 by Delmar,
a division of Thomson Learning, Inc. Thomson Learning™ is a trade-
mark used herein under license

Printed in the United States of America
1 2 3 4 5 XXX 05 04 03 02 01

For more information contact Delmar,
3 Columbia Circle, PO Box 15015,
Albany, NY 12212-5015

Or find us on the World Wide Web at http://www.delmar.com

Library of Congress Cataloging-in-Publications Data

Fryman, Mark A.
    Quality and process improvement / Mark A. Fryman
        p. cm.
    ISBN 0-7668-2872-7 (alk. paper)—ISBN 0-7668-2873-5 (alk. paper)
    1. Quality control. 2. Efficiency, Industrial. I. Title

TS156 .F776 2002
658.5'62—dc21                                              2001028284

**NOTICE TO THE READER**

Publisher does not warrant or guarantee any of the products described herein or perform any independent analysis in connection with any of the product information contained herein. Publisher does not assume, and expressly disclaims, any obligation to obtain and include information other than that provided to it by the manufacturer.

The reader is expressly warned to consider and adopt all safety precautions that might be indicated by the activities herein and to avoid all potential hazards. By following the instructions contained herein, the reader willingly assumes all risks in connection with such instructions.

The Publisher makes no representations or warranties of any kind, including but not limited to, the warranties of fitness for particular purpose or mer-chantability, nor are any such representations implied with respect to the material set forth herein, and the publisher takes no responsibility with respect to such material. The publisher shall not be liable for any special, consequential, or exemplary damages resulting, in whole or part, from the reader's use of, or reliance upon, this material.

# Contents

# Preface

Quality has always been the key to business success and long-term survivability. Quality has been synonymous with the philosophies of experts such as Dr. W. Edwards Deming, Dr. Joseph M. Juran, Dr. Genichi Taguchi, and others. Each expert possesses his own sound philosophy and his own sound approach for the improvement of quality. Amidst the experts' different approaches there has always been that one single thread that has tied all of their philosophies together. That single thread is that quality is a science—a science with very provable methods and approaches. These provable methods revolve around quantification.

In recent years, and especially with the fast-paced environment that today's businesses operate in, quality management in industry and government has not always been approached scientifically. Rather, in many instances, the hope for instant success has led to shortcut efforts in which the entire quality movement within an organization has been diluted and reduced to a "cost-savings" initiative. Time and time again, industry and government have implemented process improvement teams whose main goal is to reduce costs or produce their current product or service in a more efficient manner. And, unfortunately, sometimes the focus of improved efficiency and reduced costs have ignored two of the main ingredients in quality:

1. The customer
2. The basic understanding that quality is a system, and this system contains processes as its nucleus

More times than not, these short-sighted initiatives have resulted in short-term positive results on the profit and loss statement, but have done nothing whatsoever to guarantee the long-term survivability of the organization. After all, it is possible to produce a product efficiently and still have a product that is not purchased by customers.

It is also possible to reduce the cost of manufacture of a product and still have a product that is not purchased by customers.

The bottom line to any organization that ignores quality is that reality changes, and that the current reality of being in business today may not necessarily be the future reality of the organization. America's corporations have proven time and time again that reality is that customers may not necessarily come back and buy their products, and this reality translates into reduced market share or, sometimes, extinction.

The goal of this text is first to analyze the many interrelated concepts of quality, then to show how these concepts provide the basis for the specific step-by-step roadmap for documented process improvement and improved product quality. The roadmap steps are based on the quantification approach to quality, and the quantification integrity of the roadmap is maintained through each step.

The text is written similarly to a recipe, in that it is ready for application to the many blue-collar and white-collar processes that exist in industry and government. It is ready for the practitioner of quality to pick up and use on a daily basis in the quest for constant improvement of the processes that produce quality products and services.

In conclusion, software programs have been written for some of the concepts addressed in this text. They are included in the CD at the back of the book to assist the student and practitioner with process improvement applications. The programs are for developing a list of random numbers, the Z-test, B-bar and R charts, and P charts. The software should assist the student and practitioner in applying the principles of process improvement.

## Acknowledgments

It takes a lot of time and effort to write a book. It also takes a lot of time to prepare a book for publication. Even though there is only one name on the cover, the author is well aware that it is a team effort. It would be a gross error to omit the members of the team who were so vitally instrumental in the process of this publication. And with that, the author is deeply grateful to the brilliant team members.

First and foremost is my mother, Charlotte, whose patience and motivation made this product a reality. Dawn Daugherty, editorial assistant at Delmar Thomson Learning was always there when I had questions, and always there with the answers. Her sincerity and dedication to this text were not only a cornerstone but indicative of a class act. Mark Huth is the acquisitions editor on this project, and without him this product simply would not have gotten off the ground. His professionalism and sincerity were the initial motivation for me even agreeing to write. The expertise of Lara Stelmaszyk at Publisher's Studio has helped put the icing on the cake.

The author and Delmar wish to acknowledge and thank the individuals on the review panel for their suggestions and comments. The review panel consists of the following:

Edwin Landauer
Clackamas Community College
Oregon City, OR

David Devier
Owens Community College
Toledo, OH

Thomas C. Lavender
Catawba Valley Community College
Hickory, NC

John M. Thompson
California University of Pennsylvania
California, PA

Jerold C. Knight
Bridgeland Applied Technology Center
Logan, UT

Thomas E. Roberts
Milwaukee Area Technical College
Milwaukee, WI

Dr. Samuel C. Obi
San Jose State University
San Jose, CA

# PART

# I

# Underpinnings of Quality and Process Improvement

## Overview: Chapters 1–4

The first four chapters serve as the foundational building blocks necessary for effective understanding and application of the quality roadmap. Imagine how effective a person would be jumping into Calculus IV without understanding the fundamentals established in the prior courses of Calculus I, II, and III. Imagine how effective someone would be jumping behind the wheel of a car and driving on the highway without an understanding of the rules of the road. And finally, all too often, imagine how effective someone is when he jumps in and changes a process without understanding any of the implications or ramifications associated with quality and process improvement.

Industry and government leaders deal with processes. It is the quality of the goods and services produced by these processes and delivered to customers that result in an industry being either a success or a failure. One of the key fundamentals of quality is that it is a science with very provable methods, as is the improvement of processes that deliver products and services. It is this logic that drives home the point that a solid understanding of the fundamentals is important.

The first four chapters deal with the fundamentals of quality, measurement for quality, statistics for quality, and processes. These four chapters provide an understanding for effective application of the roadmap for quality, which is detailed in Part II.

# 1

## The Fundamentals of Quality

### 1.1  Chapter Objectives

After completing this chapter, the student should be able to:

- Understand **quality** and the key terms associated with quality.
- Identify the gurus of quality and understand the key concepts of their philosophies.
- Understand some of the guidelines and awards as well as the driving philosophies behind them.
- Understand some of the key business concepts such as inspection (especially 100% inspection), productivity, and specifications, as well as understand their relationship to quality.

### 1.2  Key Terms

American National Standards Institute (ANSI)
American Society for Quality (ASQ)
Cost of Quality
Philip Crosby
Customer Requirements
Defect
Dr. W. Edwards Deming
Detection

External Customer
Inspection
Internal Customer
International Organization for Standardization (ISO)
Dr. Joseph M. Juran
Malcolm Baldrige
Malcolm Baldrige Award

Plan/Do/Check/Act Cycle, or Plan/Do/Show/Act Cycle
Prevention
Productivity
Quality
Quality Assurance
Quality Control
Specifications
Total Quality Management

## 1.3   Introduction

The purpose of this chapter is to lay the important foundation for quality management, discuss some of the key people responsible for bringing quality management to where it is today, and examine key concepts such as productivity, specifications, and inspection and their relationship to quality management. The topics and associated theory in this chapter will be seen time and time again in the quality roadmap, and understanding of these concepts is a definite prerequisite to effective process improvement.

## 1.4   What Is Quality?

Quite simply, quality is giving the customer the right thing right the first time. Quality is not a slogan, an art, an opinion, or a feeling. It is a science with very provable methods (as will be seen in the remainder of this text). It is a science that must be embraced and implemented for long-term survival. As time and distance become irrelevant in today's global economy, it must be realized that the world is quickly becoming one marketplace, depending on which product or service a company produces.

Quality is not a program that is the responsibility of the quality department. Nor is quality a department that incurs additional costs because of manpower in the form of inspectors. Quality is the responsibility of everyone in the company and must be led by top management.

Quality is not the transparent world of slogans and buttons that claim "Quality is Job 1" or "Quality Is Our Name." Quality is not "talking the talk" but failing to "walk the talk." It is not the establishment of a **Total Quality Management (TQM)** label, because some companies that have "installed" this label have failed, while others have succeeded. In essence, it is not cute little phrases that many people in today's society enjoy hearing. Simply stated, quality is ACTION!

Quality is a system, which, when implemented, yields increased market share and reduced scrap and rework, just to name a few. Quality is the umbrella of a plethora of process improvement techniques and theories that starts with a company's vendors and extends beyond the sales of that company's products and services to the consumer. Quality is a system that is built on these provable process improvement techniques, which serve as components under the umbrella.

As discussed in Chapter 5, the words of quality (giving the customer the right thing right the first time) are translated into their mathematical definition. From that point the science of quality will be actively applied to generate documented process improvement.

Frequently, quality is viewed as an impediment to manufacturing and production. After all, some people believe that implementing quality will take time and money to stop production and fix what needs to be fixed in the process. But the real question is, "What would it cost if quality was not implemented and the process not fixed?"

**EXAMPLE 1**    It is May, and a company by the name of AmerTool Inc. makes high-precision tools and possesses several high-volume clients. Three of their large customers have put in extremely large rush orders. The orders would require nine-month production runs with shipments starting in five weeks, and continuing on a weekly basis. Furthermore, AmerTool Inc. realizes that several quality problems exist, and in the summer months, from June to August, they plan to start a plantwide, intensive process improvement initiative geared at reducing variability and improving quality.

These new rush orders have aggressively compressed AmerTool's production schedules and thrown a wrench into the timetable for implementing the process improvement initiative. AmerTool is very dedicated to production schedules, which they feel are extremely important. The orders are for tools that are produced on the lines currently experiencing the quality problems. The choice AmerTool has is clear: they can continue production and fix the problems later, or continue their pointed quality initiatives. Although the answer seems obvious, AmerTool chooses to process these rush orders for their very important high-volume clients and totally put the process improvement initiative on hold.

What AmerTool failed to realize is that their incorrect choice would ultimately make an impact on their production schedule goals. Production schedules had to slip because shipments were held up until some of the products could be reworked to meet the customers' requirements.

AmerTool made a decision to continue business as usual, oblivious to the fact that the rework generated was eating up 18.3% of their profits. The choice to ignore the integration of quality and process improvement into their daily business really did nothing but impede the company throughput to the customer in terms of delayed production schedules as well as decreased profit.

## 1.5  Basic Terms and Definitions of Quality

Some of the key terms and definitions of quality are explained in this section to set the stage for this chapter. It must be noted that a more exhaustive repository of definitions is contained in the appendix, but some of the initial definitions are explained here.

**Quality Assurance**: Every planned task and action necessary that demonstrate that the product or service satisfies the given customer requirements. It is the assurance that the product or service meets the criteria of the customer, and is similar to the auditing aspect that ensures that the financial records reported in a company's financial statement are correct.

**Quality Control**: All operational techniques necessary to satisfy all quality requirements. Inclusive in quality control is process monitoring and the elimination of root causes of unsatisfactory product or service quality performance.

NOTE: Even though the terms *quality assurance* and *quality control* sound similar (and some companies use them interchangeably), they are not the same term. Quality control has more to do with the actions on the production floor to control the quality level.

**Internal Customer**: The next person in the company who receives your product or service.

**External Customer**: The end user of the product or service.

**Customer Requirements**: Performance standards associated with specific customer needs.

**Detection**: A reactive strategy that attempts to identify and correct a faulty product or service after it has been produced.

**Prevention**: A proactive strategy that attempts to identify and correct a faulty product or service before it has been produced. For example, identification and correction during the design phase or development stage, or production monitoring and controlling of process parameters that have been proven to strongly influence output characteristics.

**Defect**: A state or condition of nonconformance to customer requirements that makes the product/service unusable.

**Inspection**: The act of measuring, checking, analyzing, examining, and testing characteristics of an item, product, or process, and comparing that result to specified requirements to determine a degree of conformity.

**Productivity**: A measure of output to input.

**Specifications**: Specific and measurable attributes that convey the customer requirements.

## 1.6 The Quality Gurus

In the world of quality, there are a few gurus whose names will forever be synonymous with quality. The main gurus who are studied in this chapter are Dr. Deming, Dr. Juran, and Philip Crosby. Dr. Taguchi, who is likewise considered to be a quality guru, is discussed at length in Chapter 12. Although the tactical approach of each varies, their underlying philosophies and theories set the stage for the science of quality as we know it today.

### 1.6.1 Dr. W. Edwards Deming

**Dr. W. Edwards Deming** was a renowned worldwide management consultant for nearly 40 years. He graduated in 1928 with a Ph.D. in physics from Yale University and worked with Walter Shewhart (designer of the control chart in 1924) at Bell Labs and the Hawthorne Plant. He is and has been a hero to Japanese industry, although not all American industry has been totally introduced to, or accepted, his philosophy. In fact, his philosophy was basically ignored by Western industry for approximately 30 years. Toward the end of his career, Dr. Deming began to travel extensively, introducing and acquainting Western industry with his philosophy. It is this philosophy that has been credited for Japan's economic success; in fact, one of the highest honors a company can win in Japan is the Deming Prize. Although many associate his name with the concept of **Statistical Process Control (SPC)**, it is his entire philosophy and its implementation that has resulted in numerous successful companies.

At the base of his philosophy are some basic concepts:

- The processes in the system create variation in the products and services that they create. Thus, everything varies.
- These variations can be segregated into **common cause variation** and **special cause variation**. Common cause and special cause variation are discussed in depth in Chapter 8.
- Management is responsible for creating the system (management sets the policy for the methods, manpower, and material of the system).

Dr. Deming's philosophy is best described with his fourteen points, which are also known as management obligations. These points must be totally embraced and endorsed by management to implement the Deming philosophy. His points are shown next, with a brief explanation following each point:

1. Create constancy of purpose toward the improvement of products and services. EXPLANATION: The purpose of an organization is to meet customer needs and requirements. This is achieved through innovation (the organization must be dedicated to innovation in every area), and focus on long-term survivability as opposed to short-term gains. The common mindset industrial concept of "Design the product, make the product, and sell the product" needs to be modified to incorporate today's and tomorrow's customer so that the company will be in business tomorrow as well as today.

2. Adopt the new philosophy. EXPLANATION: We can no longer live in an economy with commonly accepted levels of delay, poor quality products and services, and poor service. Management must develop an intolerance to these evils. Pride in the workmanship of quality

products and services and continual striving to be the best will result in long-term survival and better products and services.

3. Cease dependence on inspection and mass inspection as a means of achieving quality.
EXPLANATION: Build quality into the product or service in the first place and require statistical proof that the quality is built in. Inspection is a reactive, after-the-fact costly mode of doing business. Rather, focus on the improvement of processes that produce defects.

4. End the practice of awarding business based on price tag alone.
EXPLANATION: The criteria of price tag has no meaning without some measure of quality. Work toward building long-term relationships with single suppliers. Eliminate suppliers that cannot qualify with statistical evidence of quality.

5. Constantly and forever improve the system of production and service.
EXPLANATION: The philosophy of quality must be embraced in that every single activity must incorporate improvement, and this improvement comes from studying and understanding the processes. Likewise, most of the responsibility for the improvement lies with management.

6. Institute modern methods of training on the job.
EXPLANATION: The concepts of statistical methods and techniques must be taught to workers. In addition, the philosophy of quality work must be explained and taught to workers.

7. Institute modern methods of supervision.
EXPLANATION: Supervision should strive to provide manpower with the methods, machinery, and materials required to do a better job. Barriers or constraints that prohibit workers from doing their jobs must be identified and removed. In addition to top management commitment to quality and improvement, a communications channel from the workers to top management must be established.

8. Drive out fear in the workplace.
EXPLANATION: The philosophy of eliminating decision making that is based on opinions or politics versus information derived from data must be incorporated without fear of retribution. Likewise, workers must not be afraid to elevate problems, mention their concerns, or express their ideas.

9. Break down barriers between staff areas and departments.
EXPLANATION: The team approach must be implemented to address problems that hinder throughput. People in sales, research, design, and manufacturing must work as a team to foresee production problems that may be encountered

with various materials or specifications. These areas must learn to work together toward the optimization of the entire corporation as opposed to their local departments.

10. Eliminate slogans and numerical goals for the workforce.
    EXPLANATION: The real goal is continual and never-ending improvement in the production of goods and services. Work goals often lead to resentment by the workforce. Slogans, targets, and goals should be eliminated because the workforce has little input and control in the establishment of these targets and goals.

11. Eliminate work standards and numerical quotas. Substitute leadership instead.
    EXPLANATION: Numerical quotas are exactly that: they focus on quantity as opposed to quality. Laying the blame on workers for low quality or product defects is incorrect because workers have little control over the processes and systems that created the defects in the first place.

12. Remove barriers that hinder the hourly worker.
    EXPLANATION: Pride in workmanship is essential. Barriers that prevent the workers' right to pride of workmanship, including the elimination of annual or merit ratings, management by numbers, and management by objective, must be eliminated.

13. Institute a constant and vigorous program of education and training.
    EXPLANATION: The concept of education, training, and learning must be continuous in nature and not short term. Education and training must include knowledge of systems and statistical techniques that can be applied to processes and systems.

14. Accomplish the transformation.
    EXPLANATION: Don't "talk the talk." Rather, "walk the talk" by integrating the system to accomplish the preceding thirteen points. Create a structure in everyone that will push every day on the preceding thirteen points. The transformation must be never ending and must be the focus of all activities in the organization.

Dr. Deming's fourteen points incorporate common sense and are very logical. As part of the total management philosophy, Dr. Deming also included his twenty-one Obstacles to Quality and five Deadly Management Diseases, with most of the obstacles revolving around his fourteen points. His obstacles are listed next, along with a brief explanation, and they follow the same logic and common sense as his fourteen points:

1. Lack of constancy of purpose.
   EXPLANATION: The opposite of having a constancy of purpose to ensure that the company is in business in the future is simply not having constancy of purpose.

2. Constant mobility of top management.
   EXPLANATION: Many times mobility of top management is part of doing business. Top management rotates every couple of years or with some degree of regularity, and new directions result. Although this may be a part of business, what is this impact on maintaining corporate knowledge, especially with respect to constancy of purpose?

3. Failure of management to get involved with problems of production.
   EXPLANATION: In order for management to be more effective, they need to become involved in what is going on out on the production floor, and understand the problems that hinder throughput.

4. Obsolescent curriculum in schools of business.
   EXPLANATION: The schools of business need to strengthen the business curriculum with quality, productivity, and teamwork techniques.

5. Insulation via layers around top management.
   EXPLANATION: It is not common to have to go through layers and layers of people and wickets before even (if ever) getting an opportunity to talk to the top management. The delegation of authority must not be an excuse for insulation of top management from what is going on out on the floor.

6. Unending search for examples.
   EXPLANATION: How many times do companies need to see tons and tons of examples before even thinking of moving out on a new idea? The constant and never-ending search for examples can cause a company to lose out on opportunities.

7. The concept that "Our problems are different."
   EXPLANATION: The sole excuses of "We're different!" or "Our problems are different from those of that company!" can result not only in missing an opportunity, but in stagnation of any process or product improvement.

8. The expectation and hope for instant pudding.
   EXPLANATION: In the fast-paced business world, the words "We need results immediately!" or "We don't have time to study the problem" can be detrimental to any endeavor. It endorses jumping in before understanding of the process(es) is achieved.

9. Barriers that rob the hourly worker of pride of workmanship.
   EXPLANATION: Sometimes barriers exist that prohibit the workers from taking pride, even though they are the grassroots level of expertise at their job. As an example, think of some of the negative reactions that may occur when the hourly worker comes up with "a better idea" and tries to convey the idea to management.

10. Change can be difficult.
    EXPLANATION: The fact that an improvement or change is difficult is not an excuse for maintaining the status quo.

11. Many figures, no information.
    EXPLANATION: Charts, graphs, and figures abound, but many times the information about what to fix is lacking.

12. Poor teaching of statistics.
    EXPLANATION: As has been stated, quality is a science, and the understanding of what statistics is and what statistics tells you with regard to processes is paramount.

13. The thinking that "Our troubles lie entirely in the workforce."
    EXPLANATION: This obstacle hardly needs any clarification. Think how many times in a day the phrase "Who's fault is this?" is heard.

14. Dependence on final inspection on a mass basis.
    EXPLANATION: Inspection of almost everything produced to ensure we are producing a quality product or service as opposed to building quality into the product or service in the first place is reactive in nature and antiquality.

15. Inspection for extra high quality.
    EXPLANATION: Incentives, or rewards, are sometimes given when extra high-quality levels are achieved. This begs the question "Why can't this extra high-quality level be achieved continually as opposed to on an exception basis?"

16. Failure to use data from inspection.
    EXPLANATION: All too many times the data have been collected and are right there in front of our eyes. And all too many times, the data may not always be used, or we may just not want to believe what they are telling us.

17. The belief that "We installed control."
    EXPLANATION: This is similar to a denial mode of operation. Many times, it is believed by some people that control already does indeed exist in the current operations. This type of thinking may not always be 100 percent correct.

18. The unmanned computer.
    EXPLANATION: There are some people who truly believe that if it came out of a computer it must be right, and that whatever the computer does is correct. It needs to be remembered that humans, who are capable of mistakes, write the computer programs.

19. Purchasing departments must learn new careers.
    EXPLANATION: The vital role of purchasing is to ensure that the best materials are procured for use in the company's products. But just how many purchasing

agents know not only how the materials are used in the product but also how the materials and different grades may affect the quality of the product?

20. The belief that it is only necessary to meet specifications.
    EXPLANATION: Does just meeting the specifications ensure that the company will be in business tomorrow? Reality changes, and with this comes the prerequisite that a company continually strive for the improvement of products and services to meet that changing reality.

21. Anyone who comes to try to help us must understand all about our business.
    EXPLANATION: Outside consultants are in vogue. The assumption that because they are outside-of-the-company consultants they must know everything about our business is totally incorrect.

Dr. Deming's Five Deadly Management Diseases round out his management philosophy:

1. Lack of constancy of purpose.
   EXPLANATION: There must be a future market for products and services. If not, the alternative is extinction.

2. Short-term thinking with emphasis on short-term profits.
   EXPLANATION: Profits can only be declared once on a profit and loss statement. After that, the boards are cleared and it is time to start over again. If the company is interested in staying in business for a long period of time, then the thinking must reflect this philosophy and become more focused on the long term.

3. Annual system of rating salaried personnel.
   EXPLANATION: This concept does not foster a team concept; rather it is based on the presumption that to get ahead, someone must be promoted individually, or beat the competition.

4. Mobility of management.
   EXPLANATION: Continual turnover means the loss of corporate knowledge and changing direction, which impacts constancy of purpose.

5. Use of visible figures and ignoring unknown figures.
   EXPLANATION: The multiplying effect of an unhappy customer is unknown and proven by many companies whose customer base has shriveled up to nothing.

The Deming or Shewhart **Plan/Do/Check/Act (PDCA) Cycle** or **Plan/Do/Show/Act Cycle** (which it has been renamed to) is a very powerful and helpful methodology that can be used and followed for process improvement at any stage. It can be used at any level and is usually used in the team approach to improving quality.

Although Dr. Deming gave credit to Walter Shewhart for inventing it, the cycle became more associated with Deming because he recommended it for usage in Japan. The cycle is continuous and has four primary stages, with a fifth stage of repeating the cycle:

PLAN: This stage has to do with deciding what would be the most beneficial initiative to work on, what types of data would be needed, and what we expect to obtain from the data.

DO: This stage consists of carrying out the plans developed in the first stage.

CHECK (or SHOW): This stage consists of checking to ensure that the effects of the DO stage are doing what they are intended to do.

ACT: This stage incorporates studying and learning from what was done. Action may or may not be necessary at that particular point in time. In other words, action of incorporating the change if improvement did occur will result, or trying another improvement theory if the initial change did not work out. The concept is to develop knowledge that helps in providing a better product or service to the customer.

REPEAT: The repeat stage is not shown in the Deming cycle but is paramount. With the knowledge learned from execution of the PDCA cycle, this step endorses repeating the cycle with the newly obtained knowledge.

   The cycle is shown in Figure 1–1. Note that each stage of the cycle is of equal size. This is to show that no one stage gets more or less emphasis than any other stage.

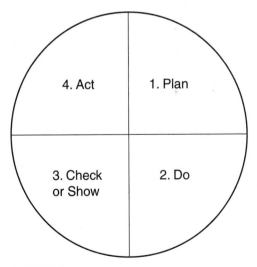

**FIGURE 1–1**
Deming Plan/Do/Check/Act or Plan/Do/Show/Act Cycle

As can be seen from the fourteen points, twenty-one obstacles to quality, and five deadly management diseases, one of Dr. Deming's fundamental building blocks is that although quality is everyone's job, management must lead the effort. His philosophy not only advocates a total system for the production of quality products and services, but sets the stage for continued growth and prosperity.

## 1.6.2 Dr. Joseph M. Juran

In addition to Dr. Deming, **Dr. Joseph M. Juran** is also credited with many contributions to the Japanese quality revolution. He was directly responsible for several high-profile successful programs in the 1980s, and three of his books (*Quality Control Handbook, Quality Planning and Analysis,* and *Management of Quality Control*) have collectively been translated into twelve different languages. He is an international lecturer and also worked with Walter Shewhart in the Hawthorne Plant in 1924. His philosophy revolved around the fact that quality represented a universal concept called fitness for use—that a vital requirement of all products and services is that they meet the needs of the society who will actually use them. In addition, the products and services must possess myriad factors that compose fitness for use, namely structural, sensory, time oriented, commercial, and ethical. The Juran Institute was founded in 1979. His philosophy can best be captured by the "Juran Trilogy," which consists of Quality Planning, Quality Control, and Quality Improvement. Dr. Juran's pillars for the trilogy include:

## Quality Planning

- Identify customers.
- Determine customer requirements.
- Develop a product that can meet the customer requirements.
- Develop processes.
- Determine process controls.

## Quality Control

- Evaluate actual performance.
- Compare actual performance to quality goals.
- Act (process improvement) on the difference between actual performance and the goals.

## Quality Improvement

- Prove the need for improvement and identify specific projects for improvement.
- Diagnose to discover causes.

- Provide remedies.
- Prove that the remedies are effective.
- Provide control to maintain the gain.

Dr. Juran was also a strong advocate for the development of accounting systems that were capable of tracking the costs of quality, with the goal of minimizing costs. He segregated the costs of quality into four categories:

1. Internal Failure Costs: these costs are defined by Dr. Juran as the costs incurred with defective products, inclusive of scrap, rework, and downtime.
2. External Failure Costs: these costs are incurred after the product has left the company and is received by the customer, inclusive of warranty costs, complaints, returned products, and rectification of the products.
3. Appraisal Costs: this classification encompasses the costs associated with finding the faulty product before it leaves the company, such as inspection and test, incoming inspection tests for vendors or supplier materials, and sustaining accurate and precise test and evaluation equipment.
4. Prevention Costs: Dr. Juran identified these costs as those resulting from keeping the defects to a minimum, including statistical process control, planning and organizing for quality, and training.

Dr. Juran's philosophy emphasized that in order to deliver quality products and services that are demanded by society, all parts of the organization must be passionately involved, from market studies and research to customer service. To further emphasize that point, when it came to quality he endorsed a position that statistical techniques were not the sole body of knowledge. Rather, it required a concentrated effort of management, statistics, and technology that coordinate together throughout the company and must be planned and organized accordingly.

Dr. Juran is also highly recognized for his "Journey from Symptom to Remedy" theory, which is shown in Figure 1–2.

## 1.6.3 Philip Crosby

**Philip Crosby** is the founder of Philip Crosby Associates, Inc. and the Crosby Quality College in Orlando, Florida. He has been a staunch advocate of quality, a highly sought-after worldwide quality consultant, and author of various quality books such as *Quality Without Tears* and *Quality Is Free*. In the 1980s he led and participated in many high-profile initiatives. He believes that the production of quality goods and services results in no additional cost to companies, but that poor-quality goods and services are what cost companies money. In addition, Philip Crosby's philosophy focused around changing management's posture toward the higher standard of performance of zero

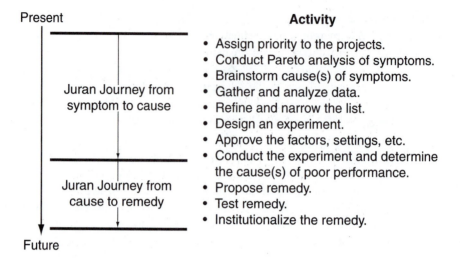

Present                                          **Activity**

Juran Journey from
symptom to cause

- Assign priority to the projects.
- Conduct Pareto analysis of symptoms.
- Brainstorm cause(s) of symptoms.
- Gather and analyze data.
- Refine and narrow the list.
- Design an experiment.
- Approve the factors, settings, etc.
- Conduct the experiment and determine
  the cause(s) of poor performance.

Juran Journey from
cause to remedy

- Propose remedy.
- Test remedy.
- Institutionalize the remedy.

Future

**FIGURE 1–2**
Juran's "Journey from Symptom to Remedy"

defects, or the never-ending search for improvement. He believed in preventive versus reactive techniques such as inspection, and he felt that management must not only shift their attitudes to prevention but communicate that this was achievable. Crosby, like Juran, endorsed quality cost measures and he created a set of "quality-building tools." One of his quality-building tools included the management maturity grid, which could be used to help isolate areas for potential improvement. His management maturity grid consists of five stages as demonstrated next:

- Stage 1: Uncertainty (no understanding or belief that quality is indeed a management tool)
- Stage 2: Awakening (the dawning that quality is important, but that is as far as it goes)
- Stage 3: Enlightenment (the establishment of a quality program within the organization occurs)
- Stage 4: Wisdom (the organization moves to prevention, with corrective action occurring on a regular basis)
- Stage 5: Certainty (the organization has now moved to a point where quality is a recognized part of doing business, with few problems occurring)

Another of Crosby's quality-building tools is his fourteen-point procedure that embeds his quality philosophy:

1. Management commitment.
   EXPLANATION: The conviction that there exists an actual need for quality is embraced by management.

2. Quality improvement teams.
   EXPLANATION: Teams are created to address the improvement.

3. Quality measures.
   EXPLANATION: Quality measures are determined for critical processes and functions within the organization.

4. Cost of quality evaluation.
   EXPLANATION: The identification of process improvement areas takes place. Likewise, initial estimates of the cost of quality are determined.

5. Quality awareness.
   EXPLANATION: Quality training to workers is accomplished with the training being conducted by the supervisors.

6. Corrective action.
   EXPLANATION: Ideas for improvement are created using information from Steps 4 and 5.

7. Zero defects planning.
   EXPLANATION: Committees create the procedures or programs plan.

8. Supervisor training.
   EXPLANATION: Training is conducted for all levels of management.

9. Zero defects day.
   EXPLANATION: Demonstration of the commitment to the new goals and standards occurs.

10. Goal setting.
    EXPLANATION: Goals are determined and not only broadcast to all areas but posted in every area as well.

11. Error cause removal.
    EXPLANATION: The causes of problems and problem areas are identified by workers and conveyed to management.

12. Recognition.
    EXPLANATION: Appreciation and recognition are conducted publicly, but the benefits are nonfinancial in nature.

13. Quality councils.
    EXPLANATION: Management and quality personnel meet on a regular basis not only to share their experiences but to share ideas.

14. Do it all over again.
    EXPLANATION: Begin again, stressing the never-ending commitment to improvement.

# 1.7 Quality System Awards and Guidelines

As with any vital science, there always exists a need for standard guidelines and awards for excellence in meeting or exceeding those guidelines. Quality is no different, and probably the two most widely known guidelines in the United States are the **International Organization for Standardization (ISO)** and the **Malcolm Baldrige Award**. The Malcolm Baldrige Award has a set of criteria, or guidelines, that need to be satisfied in order to win.

## 1.7.1 ISO

There exists in today's environment an ever-increasing trend in terms of quality performance and expectations on a worldwide basis. To address this, a series of standards has been developed. The ISO has been developing standards for international companies since approximately 1947. The ISO consists of national institutes from countries throughout the world. The ISO standards, although voluntary, set the framework and organizational guidance and requirements in various areas such as quality. The standards are not product standards or standards for Total Quality Management but system prerequisites for the company in a particular arena. For example, the ISO 9000 series (or family) is inclusive of standards that give organizational parameters and guidance on the pillars for an effective quality management system.

The American standards that are equivalent to this are the **American National Standards Institute (ANSI)** and the **American Society for Quality (ASQ)** standards. The standard, the ANSI/ASQ Q9000–Q9004, is sometimes referred to as Q9000 series. Table 1–1 depicts the ISO standard and the corresponding ANSI/ASQ standard as well as a brief description of their goal.

ISO 9001 through 9003 are actual standards, whereas 9000 and 9004 are guidelines. Figure 1–3 gives a representation of the relationship of the 9001, 9002, and 9003 standards.

**TABLE 1–1**
ISO and ANSI/ASQ Standards

| ISO Standard | ANSI/ASQ Standard | Goal of Standard |
|---|---|---|
| ISO 9000 | ANSI/ASQ Q9000-1994 | This standard is the guideline for selection and use of remaining standards in the series. |
| ISO 9001 | ANSI/ASQ Q9001-9004 | This standard details the modeling for quality assurance in the design, development, production, installation, and servicing aspects of an organization (customer interface stage, inclusive of product definition, product specification, and modification). |
| ISO 9002 | ANSI/ASQ Q9002-1994 | This standard details the modeling for quality assurance in the production, installation, and servicing aspects of the organization (inclusive of measurement, maintenance of the process, information systems, and process improvement). |
| ISO 9003 | ANSI/ASQ Q9003-1994 | This standard details the modeling for quality assurance in the final inspection and testing of products and services produced by the company (inclusive of sampling, testing, and traceability). |
| ISO 9004 | ANSI/ASQ Q9004-1994 | This standard sets the guidelines for the modeling for quality management and quality system elements and to determine the extent to which each quality system element is applicable. |

## 1.7.2 Malcolm Baldrige Award

The Malcolm Baldrige Award was established in 1987 and became Public Law 100-107. It was named after the late Secretary of Commerce **Malcolm Baldrige**, who was an advocate of quality. The American Society for Quality administers the award. ASQ is under contract to the National Institute of Standards and Technology (NIST), which is responsible for the continuation and management of the award. The Malcolm

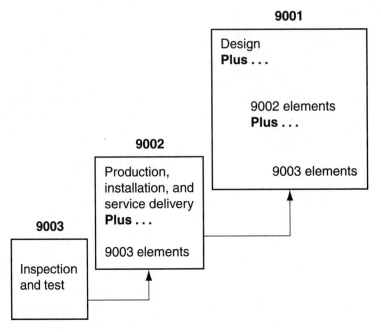

**FIGURE 1–3**
Relationship of ISO 9000 Series Standards

Baldrige Award is basically an award that promotes the awareness and importance of quality, and intends to increase U.S. companies' competitive position on a worldwide basis. The award contains criteria and guidelines for improvement in terms of more satisfied customers, increased sales, profits, and market share, and, in essence, long-term survival. Although the Baldrige Award is not as rigorous or as all-inclusive as the recommended approaches of Deming and Juran, it still is consistent with these approaches in the following areas:

- Prevention versus detection
- Meeting customer requirements
- Using statistics for process and product quality
- Process analysis and improvement
- Implementing a systematic approach to improvement

The Baldrige Award has seven categories. Each category is assigned a maximum point value as follows:

| | | |
|---|---|---|
| 1.0 | Leadership | 90 points |
| 2.0 | Information and Analysis | 75 points |
| 3.0 | Strategic Planning | 55 points |
| 4.0. | Human Resource Development and Management | 140 points |
| 5.0 | Process Management | 140 points |
| 6.0 | Business Results | 250 points |
| 7.0 | Customer Focus and Satisfaction | 250 points |
| **TOTAL** | | **1000 points** |

Each category is further broken into several examination areas to address as follow:

## 1.0 LEADERSHIP
   1.1  Senior Executive Leadership
      – Evaluates the senior leadership and personal involvement in setting direction, developing and maintaining a performance-oriented leadership system
   1.2  Leadership System and Organization
      – Assesses how the organization's customer focus and performance expectations are reflected in the leadership system as well as the ensuing management and organization
   1.3  Public Responsibility and Corporate Citizenship
      – Evaluates how the company addresses its responsibilities to the public in its performance management practices

## 2.0 INFORMATION and ANALYSIS
   2.1  Management of Information and Data
      – Evaluates the company's determination and management of information and data that are subsequently used for strategic planning, management, and overall performance
   2.2  Competitive Comparisons and Benchmarking

– Evaluates the company's processes and usage of comparison data to improve the overall performance and competitive position

2.3   Analysis and Use of Company Level Data

– Assesses how quality, customer, operational performance, and relevant financial data are analyzed and used to support company level reviews, actions, and planning

## 3.0 STRATEGIC PLANNING

3.1   Strategic Development

– Evaluates the short-term and long-term strategic planning process for competitive leadership and overall operational performance excellence

3.2   Strategic Deployment

– Assesses the development and deployment of the key business drivers

## 4.0 HUMAN RESOURCE DEVELOPMENT and MANAGEMENT

4.1   Human Resource Planning and Evaluation

– Assesses the human resource planning and evaluation as well as its alignment and integration into the strategic plan. The development and overall well-being of the workforce are also analyzed in this section.

4.2   High Performance Work Systems

– Evaluates how the company's job design and recognition programs motivate the employees to high performance

4.3   Employee Education, Training, and Development

– Evaluates how the education and training fit in with the company's plans, inclusive of growth of company capabilities and motivation

4.4   Employee Well-Being and Satisfaction

– Evaluates how the company maintains a conducive work environment and sustains the well-being and development of employees

## 5.0 PROCESS MANAGEMENT

5.1   Design and Introduction of Quality Products and Services

– Evaluates how new and improved products and services are introduced and how the processes (from manufacture to delivery) are designed to accommodate key product and service quality requirements

5.2   Process Management: Product and Service Production and Delivery

– Assesses the management of production and delivery processes to ensure quality and operational performance

5.3   Process Management: Support Services

– Assesses key support services and the management approach to ensure quality and continuous improvement

5.4   Management of Supplier Performance

– Evaluates how the company's materials, components, and other supplier-furnished services meet the company's quality requirements

## 6.0 BUSINESS RESULTS

6.1 Product and Service Quality Results
 – Evaluates the performance results of products and services using key performance measures and indicators
6.2 Company Operational and Financial Results
 – Evaluates the operational performance, financial performance, and improvement efforts using key measures and indicators
6.3 Human Resource Results
 – Assesses human resource results inclusive of the development and well-being of employees
6.4 Supplier Performance Results
 – Evaluates the results of supplier performance and process improvement initiatives using key measures and indicators

## 7.0 CUSTOMER FOCUS and SATISFACTION

7.1 Customer Market and Knowledge
 – Assesses how the company establishes short-term and long-term customer requirements and develops strategies to understand and anticipate customer needs
7.2 Customer Relationship Management
 – Evaluates management responses and follow-ups with customers in an effort to establish and build relationships, increase knowledge about their customers, improve customer performance, and generate new and improved ideas for products and services
7.3 Customer Satisfaction Determination
 – Assesses how the company determines customer satisfaction, and how their customer satisfaction compares to competitors
7.4 Customer Satisfaction Results
 – Assesses how the company measures customer satisfaction using key performance measures and indicators

Points are awarded based on a company's institutionalization of the philosophy prescribed by the award criteria. A pictorial of the Baldrige Award conceptual framework is detailed in Figure 1–4.

Although several companies can win the award in the same year in the fields of manufacturing and service, some of the Malcolm Baldrige Award winners include Motorola and Westinghouse Electric Corporation (1988), Xerox Business Products and Systems (1989), Cadillac Motor Division, IBM Rochester, and Federal Express

**FIGURE 1–4**
Malcolm Baldrige Award Conceptual Framework

Corporation (1990), Solectron Corporation (1991), Texas Instruments and AT&T Network Systems Group (1992), Eastman Chemical Company (1993), GTE Directories Corporation (1994), Corning Telecommunications Products Division (1995), ADAC Laboratories (1996), 3M Dental Products Division and Merrill Lynch Credit Corporation (1997), Boeing Airlift and Tanker Programs (1998), the Ritz-Carlton Hotel Company (1999), and KARLEE Company (2000).

### 1.7.3 The Relationship of ISO to the Baldrige Award, the Quality Gurus, and Total Quality Management

The biggest thing to remember when considering the Baldrige Award, the quality gurus, and Total Quality Management philosophies is that they all are not interchangeable. For example, the ISO standards are not a set of standards for a Total Quality Management system. They do not include the philosophy of Dr. Juran with respect to upper management involvement or the planning and management of quality. Likewise, the Malcolm Baldrige Award is not a substitute for the Deming Prize in Japan.

The best way to demonstrate the relationships is to start with the overarching philosophy of quality (or Total Quality Management as it is commonly referred to). TQM is, by definition, an integrated system for achieving customer satisfaction that involves all members of the organization and uses quantitative techniques to continually improve the processes. With this in mind, it can see that there are many elements under the arch of TQM: ISO, Dr. Juran and Dr. Deming's management philosophies, statistical quality control techniques, cooperative relationships between vendors and the companies they supply parts for, design of experiments, and the Malcolm Baldrige criteria.

## **1.8** Important Quality Related Issues

Quality does not operate in a vacuum with other business concepts. Rather, it interfaces with these concepts to become a key player in the total system and operating environment of the company. With this in mind, it is vital to have an understanding of the interface of quality with some of the typical business issues such as the concept of 100 percent inspection, productivity, specifications, and the cost of quality.

### 1.8.1 The Fallacies of 100 Percent Inspection (Proof and Exercise)

For many years the concept of 100% inspection was viewed as a mechanism for "inspecting in quality," or for bringing the product up to compliance for customer acceptance. In fact, the Germans were the benchmark experts when it came to executing this old style philosophy, as they would incorporate groups of inspectors at various key points throughout the plant to check and test products for quality. This type of thinking supports detection of defects before shipment to the customer. When looking at the true concept of quality it can be seen that the focus is, and should be, on prevention as opposed to detection. To assume that detection is a plausible means for catching all possible defects is not only incorrect (as seen in the exercise that follows), but can become costly as well. Likewise, it supports a theory of "Build it and we will bring it to spec afterwards."

The process of 100% inspection has built-in flaws that work counter to the goal of quality. For example, consider the actual process of 100% inspection and the many potential contributors to error:

- To sit and 100% inspect eight hours a day is extremely tedious and boring.
- There are interruptions such as answering the phone, which tend to make the process prone to errors.
- There are time standards for inspection ("Hurry up . . . we have to get the product to the customer"), which may or may not be realistic.
- There may be inspector bias (the old adage "Close enough for government work").
- Sales pressure ("It's good enough . . . just ship it").

Although the preceding list is not exhaustive, it does give an idea as to how prone the 100% inspection process actually is. To prove this, the following student exercise is highly recommended.

## Student Exercise 1

- Our company produces random number lists for a specific application by our customers.
- We sell our numbers in lots of 160.
- Some numbers are "good" random numbers, whereas other numbers are "bad" random numbers and cannot be used by the customer. We need to get an idea as to our scrap rate generated by the current process.
- You have been hired to 100% inspect lots of these random number lists before they go to the mailroom for mailing to the customer.
- Your instructions have been given to you by the chief inspector: "Our company goal is to sell as many lots of random numbers as possible. Thus, we cannot have lots just sitting around in the inspection department. Time studies have revealed that it only takes 3 minutes to inspect a lot of 160 numbers. Likewise, we do not endorse rework, so go through the lot once and only once. If you finish before the 3 minutes then you are doing a wonderful job!"

- The chief inspector continues: "Count the number of defective numbers in the lot and write it in the space provided. NOTE: a defective random number is one that lies between 431 and 564, inclusive. In other words, the customer does not want random numbers between 431 and 564, inclusive, to show up in any of the lots. Because the customer cannot use the lists with defective numbers, he will not pay for any list that has defective numbers in it. So we need to check all of the numbers on the lists."
- The first lot is shown in Table 1–2. The student may commence with the exercise, remembering that only 3 minutes are allowed.
- After you have finished Lot 1, the chief inspector walks in and states: "Another lot has just come in from the manufacturing department. So here is your second lot and you can begin to 100 percent inspect this new lot now."
- The second inspection lot is shown in Table 1–3. The student should commence with Lot 2 immediately, remembering that only 3 minutes is allowed for this lot as well.

**TABLE 1–2**
Student Exercise Inspection Lot 1

| | | | | | | | |
|---|---|---|---|---|---|---|---|
| 801 | 408 | 957 | 602 | 127 | 358 | 912 | 834 |
| 310 | 887 | 429 | 784 | 427 | 909 | 333 | 429 |
| 430 | 719 | 396 | 501 | 321 | 678 | 497 | 431 |
| 827 | 321 | 123 | 682 | 209 | 727 | 857 | 987 |
| 421 | 222 | 424 | 595 | 601 | 796 | 594 | 879 |
| 591 | 564 | 690 | 211 | 122 | 421 | 207 | 295 |
| 565 | 495 | 895 | 704 | 601 | 395 | 121 | 777 |
| 124 | 280 | 724 | 111 | 208 | 371 | 527 | 402 |
| 429 | 728 | 400 | 426 | 887 | 586 | 720 | 154 |
| 278 | 125 | 689 | 341 | 748 | 128 | 632 | 566 |
| 566 | 865 | 423 | 496 | 575 | 101 | 800 | 713 |
| 577 | 726 | 331 | 126 | 808 | 425 | 432 | 384 |
| 422 | 483 | 212 | 788 | 127 | 444 | 114 | 843 |
| 202 | 822 | 213 | 351 | 888 | 687 | 963 | 789 |
| 932 | 421 | 595 | 821 | 361 | 758 | 430 | 529 |
| 737 | 942 | 841 | 493 | 354 | 129 | 575 | 564 |
| 435 | 777 | 214 | 535 | 688 | 754 | 974 | 994 |
| 831 | 566 | 387 | 721 | 401 | 499 | 621 | 214 |
| 430 | 707 | 409 | 268 | 654 | 252 | 529 | 601 |
| 616 | 264 | 754 | 313 | 832 | 201 | 101 | 655 |

**TOTAL NUMBER DEFECTIVE:** _____

**TABLE 1–3**
Student Exercise Inspection Lot 2

| | | | | | | | |
|---|---|---|---|---|---|---|---|
| 750 | 357 | 906 | 652 | 177 | 323 | 875 | 862 |
| 259 | 836 | 378 | 834 | 477 | 874 | 296 | 457 |
| 667 | 668 | 445 | 551 | 371 | 643 | 460 | 459 |
| 776 | 270 | 72 | 732 | 259 | 692 | 820 | 948 |
| 370 | 171 | 373 | 645 | 651 | 761 | 557 | 907 |
| 540 | 513 | 639 | 261 | 172 | 386 | 170 | 323 |
| 514 | 444 | 844 | 754 | 651 | 360 | 84 | 805 |
| 73 | 229 | 673 | 161 | 258 | 336 | 490 | 430 |
| 378 | 677 | 349 | 476 | 937 | 551 | 683 | 182 |
| 227 | 74 | 638 | 391 | 798 | 93 | 595 | 594 |
| 515 | 814 | 372 | 546 | 625 | 66 | 763 | 741 |
| 526 | 675 | 280 | 176 | 858 | 390 | 395 | 412 |
| 371 | 432 | 161 | 838 | 177 | 409 | 77 | 871 |
| 151 | 771 | 162 | 401 | 938 | 652 | 926 | 817 |
| 881 | 231 | 544 | 871 | 411 | 723 | 393 | 557 |
| 686 | 891 | 790 | 543 | 404 | 94 | 538 | 592 |
| 384 | 726 | 163 | 585 | 738 | 719 | 937 | 924 |
| 780 | 515 | 336 | 771 | 451 | 464 | 584 | 242 |
| 379 | 656 | 358 | 318 | 704 | 217 | 492 | 629 |

**TOTAL NUMBER DEFECTIVE:** _____

— It is now time to see how well the student performed on the inspection job:

- Were 17 bad numbers in the first lot found?
- Were 26 bad numbers in the second lot found?
- Was it detected that the second lot only had 152 numbers and not the required lot size of 160?

— Think of some of the things that did not happen that also could have contributed to error (things that occur every day in the workplace):

- No interruptions were experienced (phones, the boss dropping by, etc.) in this exercise.
- Only inspected two lots—imagine doing this as a full-time job (fatigue and other factors would set in if this were a full-time job).

As can be seen from this simple example, there are many things that contribute to error when conducting 100% inspections. Therefore, if 100% inspection is intended to act as a detection mechanism, it may not always be accurate. Finally, it can be seen that 100% inspection is not the philosophy for "giving the customer the right thing right the first time." It is not a substitute for building in quality and is nothing more than a mechanism to attempt to filter, or catch, some of the errors out of production. The money spent on 100% inspection (manpower and inspection facilities) is better spent, as will be seen, on prevention and process improvement methods.

It is important to understand that inspection is not bad; rather, reliance on 100% inspection as a mechanism for building in quality in the first place is an antiquated concept.

## 1.8.2 The Linkage of Quality to Productivity

Classically, productivity has been defined as a ratio of some measure of output to some measure of input. Productivity goals have been used not only for companies as a whole, but for individual departments within a company. In breaking down the definition of productivity:

$$\text{Productivity} = \frac{\text{Output}}{\text{Input}} = \frac{\text{Number produced}}{\text{Manpower} + \text{Machiner} + \text{Materials} + \text{Methods}}$$

The numerator is obtained as products pass "pay points" located throughout the plant, and the denominator is obtained from the value of the resources expended to produce the products. The goal is to obtain as high a productivity number as possible. In further looking at the goal (increased productivity) it can be seen that there are many ways to mathematically increase the productivity ratio:

- Increase the numerator (increased production) while holding the denominator constant.
- Decrease the denominator (for example, with technological improvements, reduced manpower, etc.) while holding the numerator constant.
- Increase the numerator at a faster rate than the denominator.

Although there is nothing wrong with increasing the productivity of a department or company via these ways, the impact on the product or service due to the process change needs to be evaluated. Increasing productivity is a prime example of where the Plan/Do/Show/Act cycle should be applied to ensure expected results. Unfortunately, there are a number of companies so wound up in the fast-paced world of business that quick decisions are made in a vacuum without any planning or realization of the impacts (either positive or negative) that may occur as a result of these decisions.

In looking at the formula and concept it can be seen that there is one variable that is not accounted for: quality. Nowhere in the formula is there a variable that considers "giving the customer the right thing right the first time." Even if the numerator were replaced with the number sold as opposed to the number produced, the formula still does not incorporate a quality variable. It would just represent the number sold, without regard to the quality of that product or service.

**EXAMPLE 1**  Angelica Fraer is a manager of a major bookstore branch in the Northwest region. Her regional management is concerned and focused strictly on "the numbers," of which productivity happens to be one. She is told at her yearly appraisal that even though her store is one of the best stores in the district, her management would like her to improve her productivity by 10 percent over the next three months or else face the possibility of being replaced by another up-and-comer. Angelica feels her back is up against the wall and she begins to panic and feel that she must do something immediately, especially because she knows she has a reputation with upper management of being on the "fast track." She scans her productivity numbers over the past three months, which are portrayed in Table 1–4.

**TABLE 1–4**
Productivity Numbers for Angelica Fraer's Bookstore

| Month | Dollar Value of Books Sold | Dollar Value of Inputs | Productivity |
|---|---|---|---|
| October | 50,000 | 10,000 | 5.00 |
| November | 49,400 | 10,300 | 4.80 |
| December | 49,650 | 10,400 | 4.77 |

Angelica is aware of such techniques as the Deming Plan/Do/Show/Act cycle, but she feels herself buckling to management pressure and rationalizes that she must do something quick to stimulate "instantaneous" results. Therefore, she decides to implement several of her new ideas:

1. She lays off one clerk (reducing the manpower factor in the denominator) and spreads out the laid off clerk's workload to other employees as additional duties.
2. She begins a permanent-part-time initiative for her workers (a buzzword Angelica heard at a management conference). The concept is to replace full-time workers with permanent part-time workers in an effort to reduce the amount of benefits a company has to pay).
3. She cuts store supplies across the board by 15% per month (reducing the materials factor in the denominator).

4. She decides to lower the thermostat setting in the store by 10° in an effort to reduce energy costs consumed by the store.

5. In addition to putting her employees on permanent part-time, she decides to reduce their revenue sharing commission amounts by 2.5%.

Angelica implements her new ideas at the beginning of the calendar year and commences with business as usual. After the first quarter, Angelica reviews her productivity numbers, which are shown in Table 1–5.

**TABLE 1–5**
Productivity Numbers for Angelica Fraer's Bookstore
After Implementing Improvement Initiatives

| Month | Dollar Value of Books Sold | Dollar Value of Inputs | Productivity |
|-------|---------------------------|------------------------|--------------|
| January | 49,000 | 6100 | 8.03 |
| February | 48,500 | 5900 | 8.22 |
| March | 48,000 | 5850 | 8.20 |

Angelica briefs her regional management as to her productivity gains and they are ecstatic! Handshaking and back-patting are followed by a "break out the champagne" party, during which her regional management announces that Angelica will soon be promoted to a regional post to assume broader responsibilities.

In looking at the real bottom line of this story, it seems quite impossible that any management could possibly overlook the downward spiral in sales, but they did. Angelica and her management rationalized that the downward spiral was attributable to slow sales and a winter slump. It seems impossible that any management would not wonder about the long-term impact on the business as a result of Angelica's changes, or merely question if the spiral trends are expected to continue, but they did. Imagine the position the next manager who follows Angelica will be in. Possible low morale and slumping sales are not an environment that a new manager would like to inherit.

In summary, although productivity is extremely important, the relationship between quality and productivity must be recognized and understood. A company could realize record numbers in productivity by decreasing the factors in the denominator, while still having warehouses of built-up inventory that sits idle.

## 1.8.3 The Power Curve of Quality

There have been many conjectures that although quality is important, manufacturing and the normal day-to-day conducting of business are just as important—that quality techniques can be implemented "gradually" or "when we get a chance." It is time to prove that this concept is devastating to the organization and its long-term survival, especially in a global marketplace.

---

**EXAMPLE 1**

- Trec Inc. manufactures and sells tape recorders.
- Trec Inc. has been in business as long as its biggest competitor (Compete Inc.).
- Two years ago Compete Inc. embraced a corporate-wide quality and process improvement initiative as part of their strategic plan. The effort has been paying off and Compete Inc. has been experiencing benefits such as increased sales, reduced rework, and reduced warranty claims. Their resulting benefits are shown on the plot in Figure 1–5, where sales increased and improved efficiencies through process improvement initiatives are plotted against time.
- The executives at Trec Inc. have observed their competition and realize that they can no longer be oblivious to quality. They decide to implement the exact same quality techniques, plans, and initiatives as Compete Inc. (recall that Trec Inc. waited two years after Compete to begin their journey).

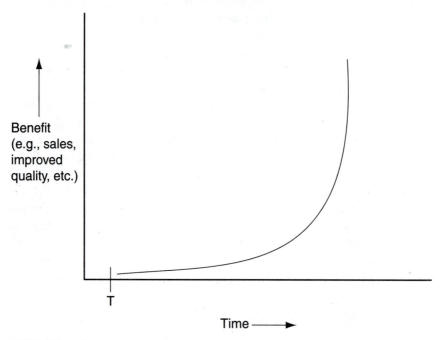

**FIGURE 1–5**
Quality Power Curve for Trec Inc.

– Trec Inc. begins to realize similar benefits over time that Compete Inc. realized. When comparing their results to Compete Inc. the chart looks similar to the one displayed in Figure 1–6.

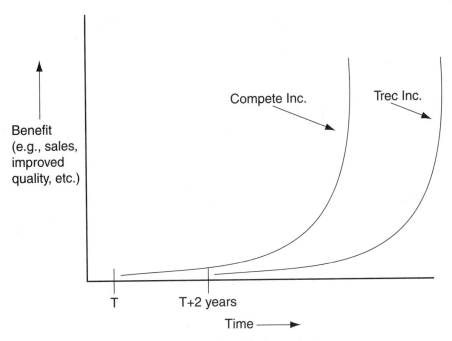

**FIGURE 1–6**
Quality Power Curve for Trec Inc. versus Compete Inc.

– On the surface, this looks pretty good as Trec enjoys tangible benefits. Some additional lines (vertical dotted lines) are put on the graph, shown in Figure 1–7, to compare the two companies.
– The added dotted lines represent the difference between the companies' sales and benefits from process improvement initiatives over time. In looking at the dotted lines from left to right (or as time continues), what do you notice about the lines? The length of the line increases. They increase as time increases. This means that as time progresses, Trec Inc. will continue to lag Compete Inc. at an increasing rate, regardless of the fact that they have replicated Compete's quality improvement initiatives. Simply stated, not only are they behind the power curve of quality, but Trec's decision to wait to implement a quality initiative after their competition had resulted in the catch-up mode of doing business. Dr. Deming's Fifth Deadly Management Disease (Search for Examples) is proven: waiting for proof and more proof that quality initiatives do work is a detrimentally negative decision.

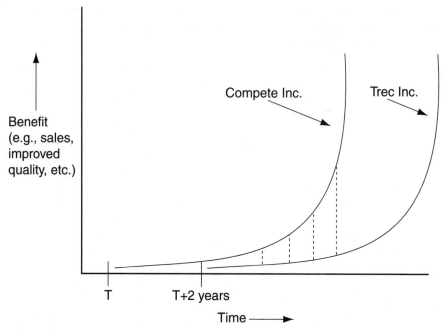

**FIGURE 1–7**
Quality Power Curve: Difference between Trec and Compete's Curves

## 1.8.4 The Concept of Specifications

Specifications are an invaluable means of converting customer requirements into product design and manufacture. Without them, a company would not know what to produce for customers. Notwithstanding, when considering quality it is very important to avoid becoming complacent with respect to specifications; to believe that specifications are perfect; or to believe that it is necessary to just meet customer specifications. There are a couple of vital concepts of process improvement when looking at the concept of specifications that prove these points:

Concept 1: Although specifications are important, it must be remembered that too many times people consider specifications in the attribute manner. In other words, "As long as it is within specification it is good." This is similar to the goalpost concept that as long as it lies between the goalposts it is good (goalpost thinking is discussed in detail in Chapter 3). This mentality is incorrect and contributes to variation and tolerance stack-up, which are all major contributors to nonquality.

Concept 2: Reference Dr. Deming's Twentieth Obstacle to Quality. The supposition that it is only necessary to meet customer specifications is antiquated. Reality changes, customer requirements change, and customers are continually demanding new and improved products and services. Consumers have an unending appetite for the very latest item on the market. Their frame of reference is always changing and their expectations are always increasing. The computer industry and clothing industry are just two examples where customers want the very latest. The specifications for what is produced today are quickly out of date tomorrow. Thus, specifications should change and should continue to change in the march to meet and exceed customer requirements. Consider how antiquated American products and technology would be if specifications were "just met" and never changed. Specifications must be viewed as vital, but not etched in stone to the point that process improvement is stalled once the specification is met.

## 1.8.5 The Cost of Quality

There are two ways to look at the **cost of quality**: (1) the actual expenses required to implement a quality culture into the organization, and (2) what costs will be incurred if quality products/services are not produced. Regardless of how the costs are categorized (Dr. Juran's four categories of cost of quality, Philip Crosby's cost of nonconformance concept, etc.) the bottom line is that if quality is not embodied in the company, then the long-term survival of the company is in jeopardy. The cost of quality is similar to viewing a mountain from a distance: the mountain top is all that is visible, and it does not look all that formidable, as shown in Figure 1–8.

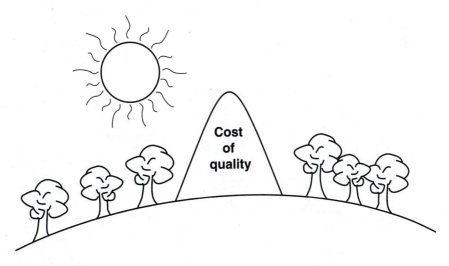

**FIGURE 1–8**
Cost of Quality Mountain from a Distance

The closer you move to the mountain, the more you begin to realize its pure size. The costs of quality are similar: from a distance the costs do not appear large, but the closer you get to them, the more apparent they are. Without categorization of the costs (and without listing all of the costs), the mountain of quality costs is shown in Figure 1–9.

It is easy to see that the costs of prevention certainly outweigh the costs of reaction to shoddy products/services.

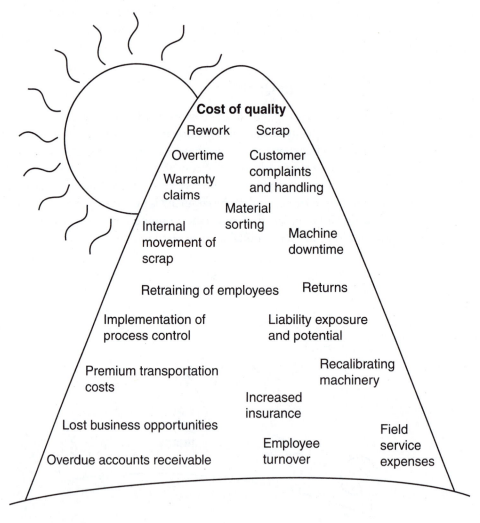

**FIGURE 1–9**
Cost of Quality Mountain Close Up

# 1.9 Quality Case I: Lands' End

Up to this point in time, the fundamentals of quality, as well as the theories of quality gurus such as Dr. Deming, Dr. Juran, and others have been explored. It is time to pay a visit in this case study to Lands' End, a company that embraces the quality concept in many avenues.

In 1963, a small company came into existence by the name of Lands' End Yacht Stores. The first location for the company was in a basement along the river in Chicago's old tannery district. Started by a group of sailors, the company's main products were racing sailboat equipment, duffle bags, rainsuits, and some miscellaneous clothing apparel such as sweaters.

As the company began to grow they relocated to a small town by the name of Dodgeville located in rural Wisconsin. As word spread of the superior quality of their clothing as well as their customer-oriented attitude, their company grew as well. The company is now worldwide and has offices in Oakham, England; Mettlach, Germany; and Yokohama, Japan, with business in almost 200 countries. In addition, they are even a leader in e-commerce, with a website at www.landsend.com.

On October 3, 1986, the company went public. Their net sales (data from Lands' End annual reports) from 1985 to 1999 are depicted in the Table 1–6.

**TABLE 1–6**
Lands' End Net Sales By Year (1985–1999)

| Year | Net Sales (dollars in Thousands) |
|------|----------------------------------|
| 1985 | 172,241 |
| 1986 | 226,575 |
| 1987 | 264,896 |
| 1988 | 335,740 |
| 1989 | 454,644 |
| 1990 | 544,850 |
| 1991 | 601,991 |
| 1992 | 683,427 |
| 1993 | 733,623 |
| 1994 | 869,975 |
| 1995 | 992,106 |
| 1996 | 1,031,548 |
| 1997 | 1,118,743 |
| 1998 | 1,263,629 |
| 1999 | 1,371,375 |

The chart of Lands' End sales is shown in Figure 1–10.

The company's results are certainly impressive with a continual upward trend every single year—in fact, the slope of the trend line for the period 1985 to 1999 is +$85,939,280.

But the clothing business is very competitive with vogue name brands abounding and abundant advertising. Although Lands' End advertises, it is definitely not a company that overindulges in advertising. So the question that begs to be answered is "Just exactly how does a company like Lands' End possess such impressive factual credentials?"

The answer lies with the company's strategic embracing of the fundamental quality principles. This strategic embracing is not documented in some book and put on the shelf but embraced in the day-to-day operations. To prove this, a sampling of examples are shown below to demonstrate the company's quality posture and attitude, not only within

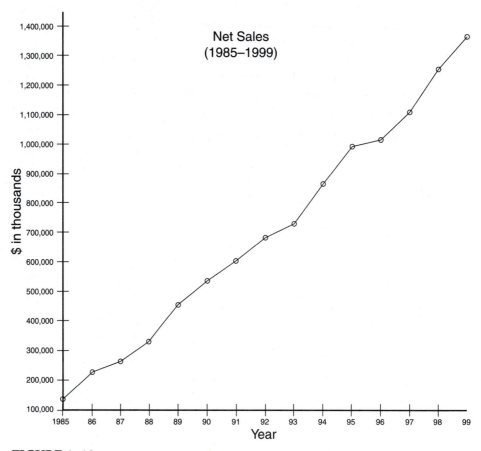

**FIGURE 1–10**
Lands' End Plot of Net Sales (1985–1999)

the walls of the company but extending to the suppliers and customers. In looking at the selected examples below, it can be seen that each one can be traced back to various segments of the quality philosophies endorsed by Dr. Deming, Dr. Juran, and Philip Crosby:

- From the 1987 annual report: "To facilitate on-site factory inspections, we recently purchased a Lear jet to provide an efficient means for our quality assurance people to reach our domestic vendors. When our goods are scheduled for manufacture, our people are at the factories to make sure our specifications are followed. The jet logged an estimated 175,000 miles this year, getting our people out to the factories to enforce our insistence on quality of manufacture."
- From the 1987 annual report: "The company devoted 64,800 hours during the year to various employee training programs. Currently, 125 Lands' End employees are receiving tuition reimbursement for courses taught by the staff of the University of Wisconsin at Platteville, which has an Extension at our facilities. Courses range from Accounting to Marketing to Comparative Statistics."
- From the 1989 annual report: "Our employees are well trained—not just when they join the company, but on a continuing basis. They are updated regularly as new products are added to the catalog."
- From the 1989 annual report: "Our biggest concern is finding suppliers who can deliver the quality and quantity needed to meet what we hope will be our continued growth. So far, we've been successful; and we believe we can keep it up. Our demands are tough, but happily we're now seeing more and more manufacturers who are able and willing to meet them."
- From the 1994 annual report: "We are installing net position inventory management programs with our largest suppliers by continuously sharing with them current sales results and future sales forecasts."
- The annual report atmosphere is almost that of a yearbook, proudly displaying the workers and their pride of workmanship. The attitude is genuine, from hand cutting fabric employees to customer service representatives to computer specialists stating, *"I work with these computers to help you get your order to you promptly. Upgrading our computer this year made it possible to continue to ship orders in 24 hours."*
- The annual reports boast time and time again that customers can always expect to talk to a customer service representative. *(As a side note, when the author contacted Lands' End and wanted to talk to someone who could give some detailed information on the company, it was within 30 seconds that the Administrator of Investor/ Financial Relations, Ms. Deborah Runde, picked up the phone and proceeded to discuss their operating principles.)*
- The business principles for Lands' End, shown in Figure 1–11, have not changed over the years, are used to train employees in the philosophy, and virtually sum up the company's quality posture. In the 1997 annual report four full pages were devoted to the principles, their explanation, and examples of their business principles in action.

## The Lands' End
# principles of doing business.

### Principle 1.

We do everything we can to make our products better. We improve material, and add back features and construction details that others have taken out over the years. We never reduce the quality of a product to make it cheaper.

### Principle 2.

We price our products fairly and honestly. We do not, have not, and will not participate in the common retailing practice of inflating mark-ups to set up a future phony "sale."

### Principle 3.

We accept any return, for any reason, at any time. Our products are guaranteed. No fine print. No arguments. We mean exactly what we say: GUARANTEED. PERIOD.®

### Principle 4.

We ship faster than anyone we know of. We ship items in stock the day after we receive the order. At the height of the last Christmas season the longest time an order was in the house was 36 hours, excepting monograms which took another 12 hours.

### Principle 5.

We believe that what is best for our customer is best for all of us. Everyone here understands that concept. Our sales and service people are trained to know our products, and to be friendly and helpful. They are urged to take all the time necessary to take care of you. We even pay for your call, for whatever reason you call.

### Principle 6.

We are able to sell at lower prices because we have eliminated middlemen; because we don't buy branded merchandise with high protected mark-ups; and because we have placed our contracts with manufacturers who have proved that they are cost conscious and efficient.

### Principle 7.

We are able to sell at lower prices because we operate efficiently. Our people are hard working, intelligent and share in the success of the company.

### Principle 8.

We are able to sell at lower prices because we support no fancy emporiums with their high overhead. Our main location is in the middle of a 40-acre cornfield in rural Wisconsin.

**FIGURE 1–11**

Lands' End: Principles of Doing Business. (© Lands' End, Inc. Used with permission.)

# 1.10  Summary

- Quality means giving the customer the right thing right the first time.
- Quality must be embraced as a business way of life for long-term survival.
- Quality is not an art but a science with very specific and provable methods.
- Some of the major gurus of quality are Dr. Edwards Deming, Dr. Joseph M. Juran, Philip Crosby, and Dr. Taguchi (who is discussed in Chapter 12). Although the specific tactics preached by each of the gurus differs, the overarching theme of each is that quality is not a novelty but a requirement for long-term survival in business.
- The Deming Plan/Do/Check/Act (PDCA) cycle, or Plan/Do/Show/Act (PDSA) cycle, is a methodology that can be used and followed for process improvement at any stage and at any level.
- The International Organization for Standardization (ISO) is an international society that develops standards that set the framework and organizational guidance and requirements in various areas such as quality. The ISO 9000 family of standards pertains to quality.
- The Malcolm Baldrige Award was established in 1987 and is geared at promoting the awareness and importance of quality, as well as to increase U.S. companies' competitive position on a worldwide basis.
- The philosophies and techniques of quality are not all necessarily interchangeable. Rather, they all exist under the umbrella of Total Quality Management.
- Dependence on 100% inspection as a mechanism for inspecting in quality supports the platform of "detecting" the defects before shipment as opposed to "preventing" the defects from occurring in the first place.
- The classic productivity ratio is a measure of output to input, where input is usually the manpower inputs, machinery inputs, and materials inputs needed to produce the output. There are several ways to increase the productivity ratio: for example, increase the output while holding the inputs constant, decrease the inputs, and increase the output at a greater rate than the inputs. When viewing productivity, care must be taken to ensure that the quality factor is accounted for in the formula.
- The power curve of quality depicts the fallacy and penalty of failing to embrace quality as a key strategic part of an organization.
- Specifications are a vital mechanism for converting customer requirements to product design and manufacture. With this in mind, companies must not become complacent or consider a goalpost mentality when addressing or considering specifications.
- The cost of quality is similar to the mountain concept in that from a distance the mountain looks significant but manageable. The closer you get, however, the more

formidable the mountain becomes. Furthermore, avoiding quality as opposed to embracing it is detrimental to both the financial and long-term survivability concerns of a company.

## Review Questions

1. Most of the responsibility for quality lies with:
   a. the CEO
   b. the quality department
   c. everyone in the company
   d. the production workers on the floor

2. Anyone in the company who receives your product or service is a(n):
   a. internal customer
   b. external customer
   c. quality inspector
   d. quality control manager

3. Productivity is:
   a. a percentage of nonconformance to customer requirements
   b. a measure of a company's market share
   c. a measure of the effectiveness of the company's management
   d. a ratio of output to input

4. Elimination of work standards and numerical quotas or goals refers to:
   a. part of Dr. Juran's Quality Control theory
   b. one of Philip Crosby's fourteen points
   c. one of Dr. Deming's fourteen points
   d. none of the above

5. The Malcolm Baldrige Award criteria place:
   a. most of the emphasis on Leadership and Business Results
   b. most of the emphasis on Customer Satisfaction and Leadership
   c. most of the emphasis on Customer Satisfaction and Business Results
   d. all of the above

6. 100% inspection focuses on:
   a. prevention
   b. correction
   c. detection
   d. a and b

7. The measure of inputs for computing productivity include all but:
   a. manpower
   b. labor efficiency
   c. machinery
   d. materials

8. A state or condition of nonconformance to customer requirements refers to:
   a. detection
   b. prevention
   c. inspection
   d. defect

9. The Quality Trilogy theory is best associated with:
   a. Deming
   b. Shewhart
   c. Crosby
   d. Juran

10. The Management Maturity grid is best associated with:
    a. Deming
    b. Shewhart
    c. Crosby
    d. Juran

11. Profound knowledge is best associated with:
    a. Deming
    b. Shewhart
    c. Crosby
    d. Juran

12. Strictly speaking, Dr. Juran's costs of quality:
    a. do not include cost of repairing defects
    b. do not include cost of avoiding defects
    c. do not include cost of monitoring defect rates
    d. do not include cost of making defects
    e. include all of the above

13. The standard that sets the guidelines for an organization whose business processes include design, development, production, installation, and servicing is:
    a. ISO 9000
    b. ISO 9001
    c. ISO 9002
    d. ISO 9003
    e. ISO 9004

14. The standard that sets the guidelines for an organization that uses inspection and testing to ensure that final products and services meet customer requirements is:
   a. ISO 9000
   b. ISO 9001
   c. ISO 9002
   d. ISO 9003
   e. ISO 9004

15. The standard that sets the guidelines for an organization that carries out all of the requirements specified in ISO 9001 with the exception of the design and development aspects is:
   a. ISO 9000
   b. ISO 9002
   c. ISO 9003
   d. ISO 9004

16. The standard that sets the guidelines for selection and use of the ISO 9000 family of standards is:
   a. ISO 9000
   b. ISO 9001
   c. ISO 9002
   d. ISO 9003
   e. ISO 9004

17. Specifications:
   a. are set in concrete and never intended to change
   b. should be reviewed quarterly
   c. should always be approved by the product engineer
   d. convert customer requirements into product design and manufacture

18. Quality improvement efforts should focus around:
   a. the company and its internal processes
   b. the company's suppliers
   c. the company's customers
   d. all of the above
   e. none of the above

19. A process improvement methodology that can be used at any level and at any stage is:
   a. Deming's fourteen points
   b. Plan/Do/Check/Act Cycle
   c. The Juran Trilogy
   d. Crosby's fourteen points

20. Retraining of employees, warranty claims, and scrap and rework are all examples of:
    a. appraisal costs
    b. prevention costs
    c. cost of quality
    d. cost of internal failures
    e. none of the above

21. Performance standards with specific customer needs are:
    a. defects
    b. customer requirements
    c. numerical quotas
    d. none of the above

22. Quality planning, quality control, and quality improvement are part of:
    a. Deming's fourteen points
    b. Crosby's fourteen points
    c. Juran's Trilogy
    d. none of the above

23. Endorsing management leadership and personal involvement in setting direction, developing and maintaining a performance-oriented leadership system appears in the Malcolm Baldrige criteria as:
    a. one of the major categories
    b. an examination area under the Leadership criteria
    c. an examination under Human Resource Development and Management
    d. none of the above

24. The _____ were very good at implementing the old style inspection philosophy by putting teams of inspectors throughout the plant.
    a. Americans
    b. Germans
    c. Italians
    d. none of the above

25. A reactive strategy that attempts to identify and correct a faulty product or service after it has been produced is called:
    a. detection
    b. prevention
    c. quality control
    d. none of the above

26. All operational techniques that are necessary to satisfy all quality requirements are called:
    a. an ISO 9000 requirement
    b. Quality Assurance
    c. Quality Control
    d. none of the above

27. How many major categories are associated with the Malcolm Baldrige Award?
    a. Seven
    b. Six
    c. Five
    d. None of the above

28. Every planned task and action necessary that demonstrates that the product or service satisfies the given customer requirements is called:
    a. an ISO 9000 requirement
    b. Quality Assurance
    c. Quality Control
    d. none of the above

29. Zero defects is best associated with:
    a. Dr. Deming
    b. Dr. Juran
    c. Philip Crosby
    d. none of the above

30. Decreasing the manpower to attain new personnel goals while maintaining and increasing manufacturing goals will definitely affect the:
    a. score received in the Malcolm Baldrige criteria of Human Resource Development and Management
    b. the number of zero defects
    c. the numerator of the productivity ratio
    d. the denominator of the productivity ratio
    e. all of the above
    f. none of the above

31. "Journey from symptom to cause" is best associated with:
    a. Dr. Deming
    b. Dr. Juran
    c. Philip Crosby
    d. none of the above

32. Select which statement is true.
    a. The Malcolm Baldrige Award can be substituted for the Deming Prize.
    b. ISO can be substituted for the Malcolm Baldrige Award.
    c. The Malcolm Baldrige Award can be substituted for Total Quality Management.
    d. None of the above.

33. When considering the Malcolm Baldrige Award, which of the following is true?
    a. The National Institute of Standards and Technology administers the award.
    b. The American Society of Quality Control administers the award.
    c. Both the National Institute of Standards and Technology and the American Society for Quality Control administer the award.
    d. None of the above.

34. Which of the following statements is true?
    a. ISO 9001 contains design elements, 9002 elements, and 9003 elements.
    b. ISO 9003 contains inspection and test elements, 9001 elements, and 9002 elements.
    c. ISO 9002 contains production, installation and service delivery elements, 9001 elements, and 9003 elements.
    d. None of the above.

35. ISO, statistical quality control techniques, and the philosophies of Deming, Juran, and Crosby are all part of:
    a. Malcolm Baldrige Award
    b. Design of Experiments
    c. Total Quality Management
    d. none of the above

# 2

# Measurement of Data for Quality

## 2.1 Chapter Objectives

After completing this chapter, the student should be able to:

- Understand the concept **measurement** and its role in quality.
- Understand some of the key concepts associated with measurement, such as discrimination, dirty data, accuracy, and precision.
- Gain an initial understanding of the relationship of accuracy and precision to quality.

## 2.2 Key Terms

| | | |
|---|---|---|
| Accuracy | Discrimination | Precision |
| Attribute Data | Measurement | Variables Data |
| Dirty Data | Metrology | |

## 2.3 Introduction

In order to make an improvement on a process, or anything for that matter, something in that process will need to be changed. This begs the question: What needs to be changed? One can guess, or hypothesize, what needs to be done and just go ahead and make a change. Or one can measure to get a more factual picture to support what is really going on as well as the impact any change will have on the process. Imagine going into your doctor's office with a complaint of stomach pains, only to hear your doctor state, without even running any tests, that he has been a physician for twenty-five years and it is his seasoned medical conclusion that you will need to have your gallbladder out immediately. You would probably take on a variety of responses, the least of which would be to run to the nearest exit. This simple example sets the stage for the importance and understanding of the world of measurement. One would expect the physician to measure and gather data before recommending a change, let alone proceeding with the change! In the same manner these same standards are applied to quality and process improvement: measurement needs to occur before making changes to the process. Likewise, because quality is a science with provable methods, data measurement is at the heart of the issue. In this chapter, measurement and the various aspects associated with it are explored in detail to serve as another fundamental building block in the roadmap to quality.

## 2.4  What Is Measurement?

*Measurement* means to proportion by some measured lot. The actual measurement may be any characteristic: length, weight, time, or some other characteristic. The lot may be any unit: inches, pounds, voltage, minutes, or some other unit. Measurement is critical to quality in that it is the basis not only for determining the quality of a process, but is also the basis by which process improvement occurs. Accurate measurement is essential because as the old saying goes, "garbage in, garbage out." There are countless examples of bad decisions made on inaccurate or incorrect measurement. In industry, bad decisions directly affect the operating profit of any department, plant, or company. Management is, and should be, concerned with accurate measurement because they are responsible not only for making decisions but for controlling the costs of the area that they are responsible for as well. If management is oblivious to whether the results of measurement are correct, then they also are oblivious to whether they will make the correct decision that will affect their organization. The accuracy and precision of measurements are critical to making sound decisions.

## 2.5  A Brief History of Measurement

The field of **metrology** is concerned with measurement, and measurement is the underlying language of science. Thus, metrology is defined as the science of measurement. In its strictest sense, metrology is concerned with the measures of length, time, and mass. The earliest accounts of measurement date back to the Phoenicians during their trade on the Mediterranean Sea when they needed to design a technique for converting the quantity of their goods to some amount of currency. In this example, the importance of measurement is seen, but now comes the next question: What standard will be used?

The first recorded historical standard was that proposed and used by the Egyptians, who were very adamant about people using the same standard for measurement. It is this belief and enforcement of measurement standardization that is credited with the success of Egyptian architecture and the pyramids. The Egyptian cubit was defined as the distance of the Pharaoh's forearm, from his elbow to his fingertip. This concept spread to other civilizations. King Henry I implemented the yard as the distance from his thumb to the tip of his nose when his arm was fully extended. Although these were good starts in establishing a standard, using the human body as a form of establishing a standard possessed some flaws. First of all, life is not a continuum, and all people, including leaders and heads of states, die. Their successor would naturally have different body measurements, which would result in implementing a new and different standard. Second, one of the causes of the French Revolution was the blatant abuse of the measurement system—many discrepancies resulted between the amounts borrowed and the amounts due to be repaid. Thus, while realizing that standards were necessary, civilization also realized that standards must be linked to something other than the ever-changing human body.

The metric system was designed in 1790 and was linked to a non-changing standard length. The meter was created as the standard and defined as one ten-millionth of the distance from the North Pole to the equator passing through Paris. Although this was an advancement over the cubit, it was soon realized that the distance of a meter needed to be replicated for more practical applications. Thus, a piece of steel the length of a meter was designed in 1799.

In 1960, a worldwide improvement was made to the metric system. The renaming of the metric system to Le Système International d'Unités, or SI, took place. Likewise, the meter was redefined to represent a wavelength of krypton 86. The length was now moved from a physical earth measurement to a wavelength.

Although the metric, or SI, system was recognized around the world, it was not the only system used. The English system was also a valid system used in the United States and England. In 1866, Congress recognized and legalized the use of the metric system but did not make it mandatory in the United States. This introduced a new problem: while the metric system was popular in Europe, it was not mandatory in the United States. Therefore, with two major systems in existence, the next logical step in the progression of standardization became the ability to convert between the two systems.

In 1893, the Mendenhall Act established the inch as 25.4000508 mm in the United States, which was still different from the English inch (25.399956 mm) and the Canadian inch (25.4 mm). In 1959, all three countries agreed to a conversion factor of 25.4 mm for the inch. Soon afterwards, the linkage of a unit of length to wavelengths eliminated differences between measurements. In addition, light waves allowed for a standard that could be applied with any measurement system.

Finally, it must be remembered that no measurement system exists in nature naturally. For example, one cannot go out into nature and find an inch lying on the ground. The systems are man made and are relative. Likewise, there is no "one best" measurement system. Both the SI and English systems have benefits. The SI system is very useful in scientific applications because every unit exists in a factor of ten (1 cm = 10 mm), and the English system is very practical for every day applications. Both have their pros and cons.

## 2.6  Why Measure and What Types of Data to Measure?

We measure for many reasons:

- To make things
- To control what we make
- To help understand what is happening in the processes, or to analyze processes
- To improve processes
- To obtain information about a process from which to make decisions
- To accept or reject incoming materials and products

We measure data and convert that data into information. There are two types of data that are addressed in this book: **attribute data** and **variables data**. Attribute data answer a yes/no, go/no-go, pass/fail type of situation. With attribute data there are only two possible outcomes. Examples of attribute data are:

- A student will either pass or fail a course.
- The part either works or does not work.
- The phone is either ringing or not ringing.
- The book is either hardback or softback.

Variables data are associated with counting or measuring items. Examples of variables data are:

- The number of students in a class
- The number of defects per lot
- The number of hardback books
- The number of eggs hatched on a chicken farm on a particular day

Understanding the types of data is critical when applying the techniques of quality, such as statistical process control. Some control charts, as will be studied in later chapters, are designed for variables data and some are designed for attribute data. Without understanding data types the wrong type of chart may be selected. If the wrong chart is selected then the wrong conclusions will be made, or in terms already mentioned: "garbage in, garbage out."

Finally, it is not only vital to understand the types of data, but it is necessary to take the time to ensure understanding of:

- Discrimination
- Dirty data
- Accurate data
- Precise data

## 2.7  Data Concepts

Data have some key concepts that must be definitely understood before just plunging into the world of data collection, data analysis, and the data interface with quality. Issues such as **discrimination**, **dirty data**, **accuracy**, and **precision** must be understood in order for quality and process improvement to be effective.

### 2.7.1 Discrimination

Discrimination, by definition, is the fineness of the measurement. The measurement of 89 in. does not mean the same thing as 89.0 in., which does not mean the same thing as 89.00 in. They all have different discriminations: 89 in. has a discrimination of 1 in., 89.0 in. has a discrimination of $\frac{1}{10}$ in., and 89.00 in. has a discrimination of $\frac{1}{100}$ in. This concept is important because when collecting data it is important to know exactly what you are measuring and the degree of discrimination required. Measurement and data collection can be costly. The instruments, manpower, and time to collect data with a discrimination of $\frac{1}{1000}$ in. can be much more costly than collecting data with a discrimination of 1 in. Care must be taken to ensure understanding of the process, as well as what discrimination will be required to gather data to measure the quality or make process improvement. Each of the following situations that require data collection for quality determination would have different discriminations:

- Testing the tensile strength of a new steel to be used in building bridges versus testing the tensile strength of rebars used to hold up a garden fence
- Testing the clarity of a $5 liquid crystal digital watch versus testing the clarity of flight indicators in an F-15
- Testing the degree of lighting needed in a knee replacement surgery versus the degree of lighting needed to illuminate a neighborhood backyard

## 2.7.2 Dirty Data

Dirty data is a term used to mean that the data do not represent what they are intended to represent. The following example depicts a situation involving dirty data.

---

**EXAMPLE 1**    A bank is interested in determining the amount of time it takes for their tellers to process customer transactions. A time and motion study is conducted during a typical week and the boundaries of the measurement are from the time the customer arrives at the window from the queue to the time the customer leaves the teller's window. The time and motion person sits in a chair at the back of the bank and collects data for a given week. The data will then be aggregated and given to the branch manager from which manning decisions will be made.

---

As can be seen from this example, the data collected may be dirty because of three underlying assumptions:

1. The assumption that the clerk has completed the previous transaction when the customer arrives at the window. (How many times have you walked up to a bank window while the teller takes a minute or two to complete the paperwork from the previous customer?)
2. Likewise, from assumption 1, the transaction may not be totally completed when a customer leaves the teller's window. (How many times do customers leave the teller's window believing they have completed the transaction, with no regard as to whether the teller has completed all of the paperwork?)
3. The assumption that every measurement collected is correct. (What if there is a surge time when there are many customers in the bank and all windows are operating at the same time? What if the time and motion person's view was blocked momentarily and an estimate was injected for the real time? What if two windows released customers within a second of each other? Would the times recorded be accurate?

The data given to the branch manager for decision making therefore may not be quite what they intended to measure.

Example 1 demonstrates the importance of clean data. There are countless times similar to the example above where data are put into the hands of management and decisions are made with no regard as to whether the data were dirty. Likewise, there are countless times when decisions are made just because data are available, with no regard as to where the data came from or what they represent. The check for dirty data in industry and government must be taken seriously when dealing with process data.

### 2.7.3 Accuracy

The textbook definition of accuracy is the adherence to a standard. Accuracy is likewise a measure of central tendency and will be studied in detail in Chapter 3. In quality, the standard is the customer requirement. By the definition, it can be seen that accuracy involves comparing what is actually occurring in the process to what is prescribed by some requirement or goal. One of the most often used techniques to demonstrate accuracy is that of an archer. In our explanation, the only goal is to penetrate the target shown in Figure 2–1.

In the first attempt, the archer took aim and fired six arrows, with the resulting pattern shown in Figure 2–2.

It can be seen that in terms of accuracy, the archer was not very accurate because all six of the arrows landed outside of the prescribed goal. In fact, we can quantifiably state that the archer was 0% accurate (zero out of six shots were in the target). The archer reloaded and fired six more arrows, with the results displayed in Figure 2–3.

The accuracy had improved somewhat, and it can be seen that the archer's accuracy can be quantified as 50% accurate (three out of six shots were in the target). Once again, the archer reloaded and shot at another target, with the results shown in Figure 2–4.

The archer improved his accuracy as now six out of the six arrows (or 100%) lie within the target.

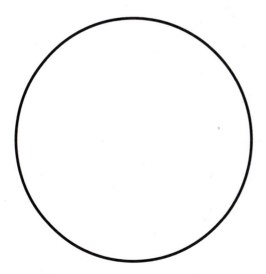

**FIGURE 2–1**
Archer's Target

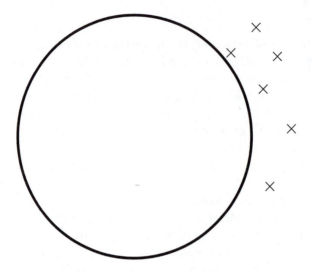

**FIGURE 2–2**
Archer's Accuracy: First Six Shots at Target

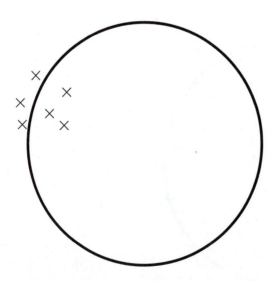

**FIGURE 2–3**
Archer's Accuracy: Second Six Shots at Target

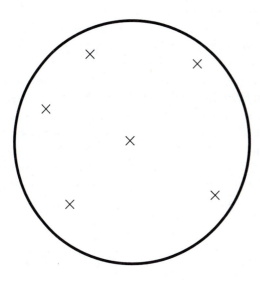

**FIGURE 2–4**
Archer's Accuracy: Third Six Shots at Target

---

**EXAMPLE 1**   A lathe setup is considered successful if less than nine minor burrs that cannot be removed in sanding occur on any table leg that is produced on the lathe. Twenty table legs were inspected with the number of burrs per leg recorded as shown in Table 2–1.

**TABLE 2–1**
Table Leg Sample: Burrs from Lathe Process

| | | | |
|---|---|---|---|
| 9 | 10 | 11 | 10 |
| 8 | 5 | 10 | 9 |
| 9 | 8 | 6 | 7 |
| 8 | 4 | 4 | 7 |
| 6 | 5 | 10 | 12 |

In terms of the prescribed goal, the lathe process is quantified as 55% accurate, because eleven out of the twenty table legs had less than nine minor burrs.

As will be seen in later chapters, the measures of central tendency (for example, mean, median, and mode) will be used to further quantify the process quality and accuracy as well as provide the means for process improvement.

**EXAMPLE 2**   Modular furniture is procured for a new multistory office building that is in the final stage of construction. The furniture wall panels are mated together with sized carriage bolts through a square (½ in. by ½ in.) cutout that lies 2 in. from the bottom of the panel to the center of the square cutout. Another mating point lies at the top of the panels in the same location from the top (2 in. from the top to the center of the square cutout) with the same square cutout size. If the cutouts are not accurate, then mating of the cubicle wall panels becomes very troublesome, time consuming, and the resulting walls will be uneven and unstable. The cutout process at the manufacturing plant has been inaccurate, specifically with the bottom cutout (2 in. from bottom to square center). Before shipping the first truckload, a sample batch of twenty is inspected with the following results shown in Table 2–2.

**TABLE 2–2**
Wall Panel Sample: Inches from Bottom to Center of Square Cutout

| | | | |
|---|---|---|---|
| 2.0 | 2.0 | 2.0 | 2.1 |
| 2.3 | 2.2 | 2.1 | 2.0 |
| 2.2 | 2.2 | 2.1 | 2.1 |
| 1.9 | 2.0 | 2.0 | 2.1 |
| 2.0 | 2.0 | 2.0 | 1.9 |

Realizing in this example that no other information is given, it can be seen that the current process can be quantified 45% (nine out of twenty) accurate.

## 2.7.4 Precision

Precision has to do with the dispersion of the measurements and is a measure of variability, which will also be studied in detail in Chapter 3. Returning to our archer example, the focus will now be on the precision aspect. Recall the three attempts at the target as shown in Figure 2–5.

In looking at the dispersion of the pattern on the first attempt, it can be seen that the arrows are not close together. Thus, the precision was not that good. On the second attempt, the precision improved because the arrows were dispersed closer together. And finally on the third attempt, the precision was worse than the first two attempts. This illustrates what precision is all about: a study of the variability of the process.

**First Attempt**

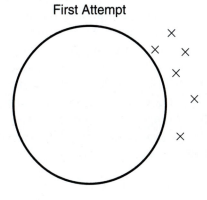

**Second Attempt**

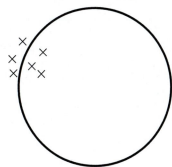

**Third Attempt**

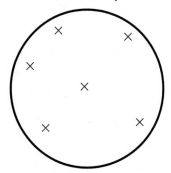

**FIGURE 2–5**
Precision: Review of Archer's Three Attempts

**EXAMPLE 1** Styrofoam cups are produced on three different production lines. The cups are intended for use in automatic coin-operated hot beverage machines. One of the production lines (Line 1) has equipment that is over twenty years old and has been experiencing significant downtime for repair due to continuous operation and tool wear. The other two production lines have three-year-old equipment that was procured from the same manufacturer. One of the engineering specifications is cup height, which should measure 6 in. from base to cup rim. The quality engineers are interested in comparing the precision of the three production lines, and the following data in Table 2–3 were collected using a discrimination of ⅒ in.

**TABLE 2–3**
Styrofoam Cup Production Line Sample

| LINE 1 | LINE 2 | LINE 3 |
|--------|--------|--------|
| 6.0 | 6.0 | 6.0 |
| 6.3 | 6.0 | 6.0 |
| 6.2 | 6.0 | 6.0 |
| 5.9 | 6.0 | 6.0 |
| 5.8 | 6.1 | 6.0 |
| 6.0 | 6.0 | 6.0 |
| 6.1 | 6.1 | 6.0 |
| 5.7 | 6.0 | 6.0 |
| 6.0 | 6.0 | 6.0 |
| 6.0 | 6.0 | 6.0 |

It can be seen that the variability, or dispersion, of the lines is different. The older production line is not as precise as either of the other two lines. When looking at the two newer lines it can be seen that while they are both more precise than Line 1, the third line is more precise than the second line.

Quantification of accuracy and precision in terms of the measures of central tendency and variability are discussed in detail in Chapter 3.

## 2.7.5 Putting Accuracy and Precision Together

Now that accuracy and precision have been looked at independently, it is time to put the two together by once again looking at the concept of target shooting. The patterns shown in Figure 2–6 depict the various combinations of accuracy and precision.

Notice that when accuracy and precision are used together, a more complete picture is given. Likewise, it can be seen that one measure is not that useful without the other.

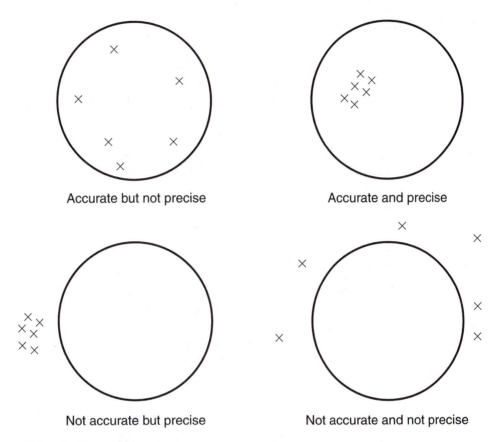

**FIGURE 2–6**
Various Combinations of Accuracy and Precision

It is the concept of precision and accuracy together that sets the stage for the measure of quality and the basis for process improvement that will be seen time and time again in later chapters.

**EXAMPLE 1**    A new design in automatic pill dispensers for pharmacies is being tested for both accuracy and precision by the manufacturer. Different compartments in the dispenser release capsules or tablets of various sizes and shapes by different desired quantities. The different medications are put in different compartments and the count is entered on the dial for the individual compartment. The new design of the dispenser would handle any size without having to be recalibrated every time new shape and different size medications are put into a compartment. A test was conducted where ten samples for three different counts of five different medications were drawn. The results are shown in Table 2–4.

**TABLE 2–4**

Pill Dispenser Sample: Counts of Various Medications

Tylenol 3 (20 count): 20, 20, 20, 19, 20, 19, 20, 20, 20, 20
Tylenol 3 (30 count): 30, 30, 30, 30, 30, 30, 30, 30, 30, 30
Tylenol 3 (50 count): 51, 51, 51, 50, 50, 50, 52, 50, 50, 50

Vicitan (20 count): 20, 20, 20, 20, 20, 20, 20, 20, 20, 20
Vicitan (30 count): 30, 30, 30, 30, 30, 30, 30, 30, 30, 30
Vicitan (50 count): 50, 50, 50, 50, 50, 50, 50, 50, 50, 50

Relafin (20 count): 20, 19, 19, 19, 20, 19, 19, 20, 20, 20
Relafin (30 count): 30, 30, 30, 30, 29, 29, 29, 29, 29, 29
Relafin (50 count): 50, 50, 50, 50, 50, 50, 50, 50, 50, 50

Aspirin (20 count): 20, 20, 20, 20, 20, 20, 20, 20, 20, 20
Aspirin (30 count): 30, 30, 30, 30, 30, 30, 30, 30, 30, 30
Aspirin (50 count): 50, 50, 50, 50, 50, 50, 50, 50, 50, 50

Ziac (20 count): 24, 24, 24, 24, 24, 24, 24, 24, 24, 24
Ziac (30 count): 33, 33, 33, 33, 33, 33, 33, 33, 33, 33
Ziac (50 count): 56, 58, 59, 58, 57, 57, 57, 57, 56, 55

The results of the prototype dispenser design can be summarized in terms of accuracy and precision as shown in Table 2–5.

It can clearly be seen from this example that both accuracy and precision are needed not only to assess the totality of the situation, but to combine to give invaluable information about the situation for which the data were collected:

- The compartments are not consistently accurate across at the 20-, 30-, or 50-count level for all of the various medications.
- Three of the four combinations of the accuracy and precision mix occurred:

  1. Accurate and precise
  2. Not accurate and not precise
  3. Not accurate but precise

This example sets the stage for the analysis of quality in processes and products, and as will be seen in later chapters, deeper quantification of accuracy and precision will provide the mechanism for process improvement.

**TABLE 2–5**
Pill Dispenser Sample: Accuracy and Precision Summary

| Medication | Count | Accurate | Precise |
|---|---|---|---|
| Tylenol 3 | 20 | 80% | No—variation exists |
| Tylenol 3 | 30 | 100% | Yes |
| Tylenol 3 | 50 | 60% | No—variation exists |
| Vicitan | 20 | 100% | Yes |
| Vicitan | 30 | 100% | Yes |
| Vicitan | 50 | 100% | Yes |
| Relafin | 20 | 50% | No—variation exists |
| Relafin | 30 | 40% | No—variation exists |
| Relafin | 50 | 100% | Yes |
| Aspirin | 20 | 100% | Yes |
| Aspirin | 30 | 100% | Yes |
| Aspirin | 50 | 100% | Yes |
| Ziac | 20 | 0% | Yes |
| Ziac | 30 | 0% | Yes |
| Ziac | 50 | 0% | No—variation exists |

# 2.8 Summary

- Measurement means to proportion by some measured lot. Data are measured and converted into information.
- We measure data for many reasons—from making things to understanding what is happening in a process.
- There are two types of data: attribute data (data that measure a go/no-go, pass/fail, yes/no type of situation, and number bad per number examined); and variables data (data that are associated with measuring items)
- Discrimination is defined as the fineness of the measurement. For example, 55.01 feet has a discrimination of one one-hundredth of a foot.
- Dirty data is a term used to mean that the data collected do not represent what they are intended to represent. Data integrity is paramount to quality and process improvement, and it is vital to ensure that data are not dirty. Rather, it is desirable for the data to truly represent the situation they are intended to depict.

- Accuracy is the adherence to some standard. Mathematically speaking, it is a measure of central tendency.
- Precision has to do with the dispersion of the measurements. Mathematically speaking, it is a measure of variability.
- Accuracy and precision are used together to describe in totality what is happening with the process.
- Both accuracy and precision can be applied to data (data accuracy and data precision) and processes (process accuracy and process precision).
- The mathematical concepts of accuracy and precision will be studied in detail in Chapter 3, and will then be addressed as measures of central tendency and variability.

## Review Questions

1. Precision is:
   a. a measure of central tendency
   b. a technique for minimizing the variability of a process
   c. the dispersion of measurements
   d. a measure of variability

2. Fineness of measurement is the definition of:
   a. precision
   b. accuracy
   c. discrimination
   d. none of the above

3. The science of measurement is defined as:
   a. metrology
   b. process improvement
   c. accuracy
   d. precision

4. The order of progression in the history of standards development is:
   a. wavelength, cubit, a portion of the distance from the North Pole to the equator
   b. cubit, a portion of the distance from the North Pole to the equator, wavelength
   c. cubit, wavelength, a portion of the distance from the North Pole to the equator
   d. none of the above

5. Data that may not always represent what they are intended to represent describes:
   a. inaccurate data
   b. nondiscriminating data
   c. imprecise data
   d. dirty data

6. Which of the following is not a reason for measuring?
   a. To understand what is going on with a process
   b. To make process improvement
   c. To accept or reject incoming material
   d. To control what is manufactured
   e. None of the above

7. The term *prescribed goal* has to do with:
   a. accuracy
   b. precision
   c. dirty data
   d. discrimination

8. What was one of the biggest contributions of the Mendenhall Act?
   a. It made usage of the metric system mandatory in the United States.
   b. It established a conversion between the metric and English systems.
   c. It established a common conversion between the United States, Canadian, and English measurement for an inch.
   d. None of the above.

9. Attribute data are:
   a. used when counting items
   b. a measure of variability
   c. a measure of central tendency
   d. used to answer questions when only two choices exist

10. The cubit was defined as:
    a. the distance between the knuckles and elbow of the Pharoah
    b. two times the distance between the knuckles and the elbow of the Pharoah
    c. the distance between the Pharoah's nose and tip of his finger
    d. the distance between the Pharoah's elbow and fingertip

11. Management decisions should be made from data that:
    a. are both accurate and precise
    b. contain no dirty data
    c. have a discrimination of one-tenth
    d. all of the above
    e. a and b

12. Even if only one data point is collected:
    a. precision can still be calculated
    b. data discrimination can be determined
    c. conversion between the English and SI systems can occur
    d. b and c
    e. none of the above

13. Counting items best describes:
    a. attribute data
    b. variables data
    c. dirty data
    d. accuracy

14. The measurement system that is excellent for scientific applications is:
    a. the SI system
    b. the English system
    c. both systems
    d. neither system

15. Proportion by some measured lot is a definition of:
    a. accuracy
    b. attribute data
    c. measurement
    d. precision

16. The Egyptians strived to:
    a. standardize measurements
    b. enforce a standard other than the cubit
    c. realize the importance of accuracy and precision
    d. none of the above

17. Invalid assumptions is a key contributor to:
    a. nonstandard measurements
    b. using variables data over attribute data
    c. dirty data
    d. improper determination of discrimination

18. In its strictest definition, metrology is concerned with the measurement of:
    a. length, time, and movement
    b. mass, weight, and length
    c. mass, time, and length
    d. none of the above

19. It is imperative in quality that data be:
    a. accurate
    b. precise
    c. clean
    d. all of the above
    e. a and b

20. "Seven tires" is an example of:
    a. attribute data
    b. data that may be dirty
    c. precise data
    d. variables data

21. When was the metric system designed?
    a. 1780
    b. 1790
    c. 1800
    d. None of the above

22. The measurement system that has the advantage in terms of practicality is:
    a. the English system
    b. the SI system
    c. both systems
    d. neither system

23. In 1959, the United States, Canada, and England agreed on the extremely important measurement concept that:
    a. the SI system would be used in scientific applications
    b. the English system would be used in scientific applications
    c. 25.4 mm/in. would be the conversion factor between the SI and English systems
    d. none of the above

24. Precision is measured with what mathematical concept?
    a. Central tendency
    b. Variability
    c. Reliability
    d. None of the above

25. Accuracy is measured with what mathematical concept?
    a. Central tendency
    b. Variability
    c. Reliability
    d. None of the above

# Problems

1. Describe two situations different from those cited in the text where:
   a. the data collected for process improvement would have a discrimination of one
   b. the data collected for process improvement would have a discrimination of $\frac{1}{10}$ in.

2. Two-by-fours are needed in the new construction of twenty homes. Three companies produce two-by-fours and are bidding for the contract to supply wood. Although all three companies are competitive in terms of price, the builder wants a sample of five from each bidder. The builder tells the bidders that the critical parameter he is interested in is going to be how close the samples are to the requirement overall length of 5 ft. Using the following data in Table 2–6 that was provided by the bidders, place the bidders in order in terms of accuracy and precision and make your recommendation as to which bidder should be awarded the contract.

**TABLE 2–6**
Bidder Data for Two-by-Fours

| BIDDER 1 | BIDDER 2 | BIDDER 3 |
|----------|----------|----------|
| 5.0 ft.  | 5.0 ft.  | 4.9 ft.  |
| 5.1 ft.  | 5.0 ft.  | 4.9 ft.  |
| 5.0 ft.  | 5.2 ft.  | 4.9 ft.  |
| 5.0 ft.  | 4.9 ft.  | 4.9 ft.  |
| 4.9 ft.  | 5.0 ft.  | 5.0 ft.  |

3. Two pulleys are tested to determine the maximum weight they can support. The data, measured in pounds, of the test results are shown in Table 2–7.

**TABLE 2–7**
Pulley Test Weight Results

| PULLEY 1 | PULLEY 2 |
|----------|----------|
| 310      | 305      |
| 295      | 315      |
| 300      | 315      |
| 310      | 290      |
| 305      | 290      |

| 310 | 315 |
|-----|-----|
| 300 | 315 |
| 296 | 315 |
| 305 | 290 |
| 310 | 295 |

Discuss which pulley would be best in terms of accuracy and precision to satisfy a customer requirement of 305.

4. Using a scenario other than the one of targets that is used in the chapter, demonstrate (with pictures or words) the concept of accuracy and precision.

5. Give five examples of variables data associated with:

| *A Manufacturing Process* | *A Service Process* |
|---------------------------|---------------------|
| 1. _____ | 1. _____ |
| 2. _____ | 2. _____ |
| 3. _____ | 3. _____ |
| 4. _____ | 4. _____ |
| 5. _____ | 5. _____ |

6. Give five examples of attribute data associated with:

| *A Manufacturing Process* | *A Service Process* |
|---------------------------|---------------------|
| 1. _____ | 1. _____ |
| 2. _____ | 2. _____ |
| 3. _____ | 3. _____ |
| 4. _____ | 4. _____ |
| 5. _____ | 5. _____ |

7. Match each of the following statements with the measurement concept it best describes:

____ "Measurement to .001 in. are needed"

____ "We need to gather data to find out why our rework is so high."

____ "Out of fifteen batteries, three failed and twelve passed."

____ "There are fourteen carrots"

____ "The data gathered do not represent the actual time to process one transaction. They represent the average number of transactions the department handles per hour."

____ "My goal was to shoot one arrow and make a bullseye. That is exactly what I did."

a. example of attribute data
b. example of dirty data
c. example of variables data
d. example of discrimination
e. example of accuracy
f. example of why data need to be collected

8. Give an example for each of the following:
   a. Describe a situation where the data collected turn out to be accurate but not precise.
   b. Describe a different situation where the data collected turn out to be precise but not accurate.

9. Describe a situation at your place of employment or home where:
   a. data need to be collected and why
   b. what data will be collected and why
   c. will the data be attributes or variables, and why
   d. what discrimination you will use
   e. how you will know (after data collection) if the data are accurate and precise

10. Give an example for each of the following:
    a. where collecting data using the SI system is more beneficial than the English system
    b. where collecting data using the English system is more beneficial than the SI system

# 3

# The Fundamentals of Statistics for Quality

## 3.1 Chapter Objectives

After completing this chapter, the student should be able to:
- Understand the basic concepts of random numbers.
- Understand the basic concepts of sampling and surveys.
- Understand the normal curve and the mathematical concepts of central tendency and variability.
- Understand the Z-statistic and how to use it to determine areas under the curve.
- Understand the theory behind basic and conditional probability and their relationship to quality.

## 3.2 Key Terms

| | | |
|---|---|---|
| Arithmetic Average | Median | Sample |
| Bell-Shaped Curve | Mode | Standard Deviation |
| Bias | Normal Curve | Survey |
| Bimodality | Probability | Tolerance Stack-Up |
| Confidence Level | Questionnaire | Z-test |
| Margin of Error | Random Sample | |
| Mean | Range | |

## 3.3 Introduction

As has been stated, and restated several times throughout this text, quality is not an art but a science with very provable methods. Those provable methods require an understanding of some of the key concepts of statistics. To date, we have looked at the fundamental philosophy of quality and the concept of measurement. In this chapter we will look at some of the statistical techniques used in quality and process improvement. These techniques convert the measured data into information from which effective decisions can be made. Random numbers, sampling, and basic statistics and probability are concepts studied in this chapter. An understanding of these statistical concepts is also a fundamental building block.

## 3.4 Basic Concepts of Random Numbers and Sampling

A **sample** is defined as a representative subgroup of a population. We sample to obtain information about the population without having to poll the entire population. An example is surveying a sample of potential customers to determine what changes need to be made to a particular product in order for them to buy the product on a continuing basis. It is not feasible, economical, or practical to poll the entire population of potential customers, so a sample is taken. As another example, every election year a sample of a state's precincts are given exit polls to determine which party will carry the state. When the returns are watched on television, it can be seen that sometimes as low as 2 percent of the precincts collected in a state are sampled to determine which candidate will win the electoral votes in the state. This is another classic example of determining a characteristic of an entire population via sampling. The key concepts of sampling are:

- The goal of a sample is to measure some characteristics from that sample in order to depict what is happening with the entire population. Thus, effective planning of the sample and sample size must take place.
- The sample needs to be random. A random sample occurs when all items in the given population have an equal probability of being selected. The only way to develop a true **random sample** is using a random number generator or a table of random numbers. You, your boss, a statistician, or I or anyone else cannot develop a random sample. Human beings possess bias: they favor particular colors, numbers, and opinions. To assume that these inherent tendencies could be overcome to obtain a pure random sample is impossible. Likewise, the mere definition of random (all items have an equal probability of being selected) implies that something more sophisticated than human selection needs to occur.

The importance of a purely random selection is vital. If a sample is not random, then the **confidence level** and **margin of error** associated with the sample will be negatively affected. For example, let us suppose that a random sample was taken from a lot of incoming bars and the length was measured. After the random sample was taken the statement made was: "We are 95 percent confident that the length of the bars in the entire lot lies between 7.3 inches ± 0.2 inch." If the sample were not random then both the confidence level (95 percent) and margin of error (± 0.2 inch) would not be viable.

- The random sample size is paramount, which leads to two very valuable statistics associated with sampling and sample sizes: confidence level and margin of error:

  1. Confidence Level: The certainty level that the measured probability of some assertion is true. Putting it another way, confidence level is the degree of belief that the information obtained from the sample represents, or is indicative, of the entire population. Although the choice of a confidence level is subjective and depends on the criticality of the project or analyst, a typical confidence level is 95%.
  2. Margin of Error: A sample is a quantity less than 100% of the entire population. In thinking about this logically, it is realized that without conducting a 100% sample, there will be some margin of error that exists. As an example, suppose a sample out of 750 yielded that 42% of the respondents felt that the cost for local public transportation was too high. The estimate of 42% does not represent the absolute actual percentage of the entire population. The actual percentage may be 42%, higher than 42%, or lower than 42%. But we would expect the estimate for the entire population to be within a ± range around 42%. This range is called the margin of error. In our example, the margin of error may turn out to be ± 2%, or from 40% to 44%.

     A quick, rough estimate for computing the margin of error is:

$$\text{Margin of error} = \frac{1}{\text{the square root of the sample size}}$$

- The data must be valid and reliable. One can sample all one wants and obtain all of the measurements, but unless the data that are being collected are both valid and reliable, then the results will be meaningless.
- To conduct a random sample, the following step-by-step procedure is followed. (Note: this is not the only technique to determine a random sample, but the concept of random sampling is effectively demonstrated.):

STEP 1: Assign a number (1, 2, 3 . . . ) to each of the items from which the sample will be drawn.

STEP 2: Determine the total number of items from which the sample will be drawn.

STEP 3: Although there are many ways to determine a set of random numbers, the use of the table of random numbers (Table 1 in the appendix is a list of 500 five-digit random numbers of uniform distribution) will be demonstrated. Determine a row and column randomly (for example, one way would be rolling two dice twice: once for the row and once for the column).

STEP 4: Using the five-digit number that corresponds with the row and column, divide the random number by 100,000 or put a decimal point before the first digit of the four-digit number (for example, if the number in the table was 26207, then putting a decimal in front of it would yield .26207).

STEP 5: Multiply this decimal by the total number of items from which the sample will be drawn (from STEP 2).

STEP 6: Truncate the decimals.

STEP 7: This new number will represent the number of the item (see STEP 1) to be placed in the sample.

STEP 8: Repeat the process for STEPS 3–7 until the total sample size has been selected. Note that there may be occasions where the same item is selected. In these cases another item must be selected because it is not permissible to use the same item more than once in the sample. This would definitely make an impact on the confidence level and margin of error.

---

**EXAMPLE 1**   It is determined that a sample of six unique items needs to be taken from a population of ninety incoming rubber grommets in a lot. Instead of rolling dice, a more sophisticated technique (a computer program using a random number seed) was used, which yielded the following random numbers:

1164, 7110, 5607, 3786, 0782, 7110, 0388—Note that seven random numbers were selected. This is because the number 7110 came up twice and it is incorrect in this particular situation to sample with a replacement (where the item is inspected, put back into the lot, and is eligible to be selected again at random).

From the above random numbers, the six items selected are as follows:

1st item: $(.1164)(90) = 10.476 = 10$

2nd item: $(.7110)(90) = 63.99 = 63$

3rd item: $(.5607)(90) = 50.463 = 50$

4th item: $(.3786)(90) = 34.074 = 34$

5th item: $(.0782)(90) = 7.038 = 7$

6th item: $(.0388)(90) = 3.492 = 3$

---

In conclusion, the linkage between sampling and surveys to quality can be discussed. When it is determined that the quality of a product or service needs to be determined, it is logical to obtain a random sample of the product or service for measurement purposes as opposed to measuring the entire population. From this sample the quality of the product or service can then be quantified, as will be seen in Chapter 5.

Also, in setting the stage for the next section, there will be times when it is desirable for marketing and product research to obtain a feel for what the customer thinks of the quality of the product or service. And the best way to do this is to take a random sample of customers and survey them as to the different product or service performance and characteristics. It may be desirable to sample people who do not purchase the product or service to obtain their views as well. This is why the concept of random samples and surveys (which is discussed in detail in the next section) is so important in the world of quality. It is through the concepts of sampling and surveys that allow for the collection of valuable information that can be used to drive decisions about the processes that deliver quality products and services.

## 3.5 A Word about Surveys

Now that a basic understanding of random sampling has been achieved, focus can shift to collection of data through surveys. There are many different techniques for collecting data, one of which is a **survey**. Surveys are an excellent instrument for collecting data that may not be already available, or for data that cannot be easily collected. They are used to focus on some particular problem, and can be conducted by either interviewing people or by using a questionnaire. It must be remembered that conducting interviews is much more time consuming, respondents may or may not always be totally at ease with the face-to-face situation, and accurate and precise information may not always occur. Some of the advantages of surveys as a technique for collecting data include:

- The survey can reach a large population simultaneously.
- The cost of surveys can be relatively inexpensive compared to some of the other data collection techniques.

- The survey offers the respondents (the people taking the survey) a mechanism for not only answering at their own convenience (within the survey timeframe), but for expressing their feelings without fear of embarrassment or reprisal.
- The data are usually easy to handle and tabulate.

As with the determination of a random sample, there are steps to follow when developing and conducting a survey. The first step in conducting a survey is to identify the problem or focus area. It is important to know specifically what the area of focus the survey is concerned with and is seeking information about. After identifying the problem, the next step is to conduct a very thorough literary research to investigate what has already been published on the particular topic. This step is not only beneficial in determining whether the survey needs to be conducted, but gives insight into questions to be asked. As can be seen in Figure 3–1, the first two steps are also sequential.

After the first two steps have been conducted, it is time to begin predevelopment of the **questionnaire**. A questionnaire is a form for eliciting and recording data. The predevelopment stage of the questionnaire involves following some basic guidelines that will not only optimize the information obtained from the results, but will increase the efficiency of the survey. For example, there is nothing more discouraging than to completely conduct a survey only to realize that the right questions were not asked, or the questions that were asked will not achieve the objective. The survey guidelines are:

- Map out the parts of the survey (for example, instructions, timeframe for completion).
- Generate a hypothesis about the data that are being collected; naturally, the more information that can be obtained from the data the better.
- Establish decision rules for accepting or rejecting each hypothesis (for example, statistical test of significance).
- Determine the appropriate sample size that will yield the desired confidence level and margin of error.
- Determine how the data will be recorded and collected.
- Determine what methodologies for analysis of the data will be performed on the data when they are collected (for example, quantitative analysis, qualitative analysis, or quantitative-qualitative analysis).

**Survey step 1:** Identify the problem area or area of focus.

**Survey step 2:** Conduct a thorough literary research.

**FIGURE 3–1**
Survey Steps 1 and 2

The first three sequential survey preparation steps are shown in Figure 3–2.

By now, after the predevelopment stage of the survey, one thing should be very apparent: planning is imperative to a successful survey. The probability of conducting a successful survey without planning is highly unlikely.

The next stage in surveys is the development of the questions for the questionnaire, and planning will continue to play an important role in this step as well. The development of the questions is very critical and not as easy as it sounds. Many questions that appear on the surface to be good survey questions contain **bias** that can skew the results of the survey. Bias occurs when the replies of respondents are influenced in a one-sided manner by the way the survey questions are worded. In 1951, Stanley L. Payne analyzed what sample researchers saw as the principal problems with research methods. The results, reported in *The Art of Asking Questions* published by Princeton University Press, are shown in columnar format:

| | |
|---|---|
| Improperly worded questions | 74% |
| Faulty interpretation | 58% |
| Inadequacy of samples | 52% |
| Improper statistical methods | 44% |
| Presentation of results without supporting data | 41% |

**Survey step 1:** Identify the problem area or area of focus.

**Survey step 2:** Conduct a thorough literary research.

**Survey step 3:** Predevelopment stage of survey inclusive of:
- Mapping out the parts of the survey, including instructions
- Generating hypothesis about the data being collected
- Establishing decision rules for accepting or rejecting the hypothesis
- Determining the sample size and an appropriate confidence level and margin of error
- Determining the data collection and recording techniques
- Determining analysis methodologies

**FIGURE 3–2**
Survey Steps 1 through 3

Although the age of the report may cause one to question its accuracy in the current fast-paced technologically efficient society of the late twentieth and twenty-first centuries, a simple question will help validate its accuracy. Put a check mark in the appropriate response:

\_\_\_ I am in the upper class of society in the United States.

\_\_\_ I am in the middle class of society in the United States.

\_\_\_ I am in the lower class of society in the United States.

The above question looks like it is a fair question, but it has bias in it. First of all, it includes stereotype answers that are associated with the different classes of society, and some of the results may be considered undesirable. As another example, check the appropriate response to the question below:

\_\_\_ I smoke

\_\_\_ I do not smoke

What was your initial thought after reading the question? Chances are it was not about answering the question on the survey. In today's society where it is very vogue to protect people's rights, a politically incorrect answer (for example, I smoke) to this question could be more dangerous to one's health than the actual act of lighting up.

A checklist of items to consider when developing questions for a questionnaire include:

- Who is asking the question (especially if it is an oral interview)?
- What kind of information must be obtained from the answer?
- Will there be a decision based on the information?
- Who will answer the question?
- What will occur if incorrect answers are given? (For example, will the wrong decision be made if false answers are given?)
- What type of questions will be used?

  > Open-ended: a very easy to ask question in which the respondents' knowledge is sought; for example, "What is the best way to structure this company and its subsidiaries?"

> Multiple choice: a question that has the only options spelled out in the respondent choices; for example, "The most important factor in determining the location for a family residence is: (a) Affordability, (b) Location, (c) Family Agreement, (d) Convenience."
> Two-way: a question that allows as its answers one of only two choices, such as true or false or yes or no; for example, "I agree that this product is heads and tails above the competition." Yes ___ No ___

Note: Each type of question has pros and cons that the researcher should be aware of before selecting that particular type of question. For example, with a two-way question a definite pro is that the respondent has only one of two options, whereas a con is that two options may not be enough because the answer may not be black or white.

• Review the questions to ensure that the following questionnaire pitfalls have been avoided:

> Does it appear that the questions could be misinterpreted?
> Does the question have stereotypical answers in it?
> Are you making any assumptions regarding the respondent's knowledge of the subject matter?
> Do any of the questions "lead" the respondent to answer in a particular manner?
> Does the question use grammar and expressions that the respondent may not be familar with?
> Does the question have answers that are socially unacceptable if answered in a particular manner?
> Is the question too long, possibly to the point that misinterpretation could occur?
> Is there value to asking the question or is the question really needed to meet the objectives of the survey?

The four steps for preparing the survey can now be seen in totality in Figure 3–3.

Although the preceding checklist is not an exhaustive discussion on surveys, it does point out some of the key considerations when constructing a survey to collect data. The old saying that "garbage in equals garbage out" seems like it was almost written with surveys in mind. Violation of any one or a combination of the points could result in "dirty data" (or garbage in), which will definitely result in "garbage out." And when it comes to gathering customer data regarding the quality of the product or service, it is imperative that the guidelines of surveys be adhered to. It could be detrimental to a company to make changes to a product or service based on the results of a bad survey.

**Survey step 1:** Identify the problem area or area of focus.

**Survey step 2:** Conduct a thorough literary research.

**Survey step 3:** Predevelopment stage of survey inclusive of:
- Mapping out the parts of the survey, including instructions
- Generating hypothesis about the data being collected
- Establishing decision rules for accepting or rejecting the hypothesis
- Determining the sample size and an appropriate confidence level and margin of error
- Determining the data collection and recording techniques
- Determining analysis methodologies

**Survey step 4:** Develop questions using a checklist inclusive of:
- Who will be asking the question? What kind of information must be obtained?
- What type of decision will be made? What will occur if incorrect answers happen?
- What type of questions will be used? Open-ended? Multiple choice? Two-way?
- Test the questions for pitfalls such as:
  - —leading questions   —possible misinterpretation of questions
  - —biased questions    —question clarity
  - —length of question  —is the question necessary and of value
  - —confusing grammar

**FIGURE 3–3**
Survey Steps 1 through 4

# 3.6 The Normal Distribution

The normal distribution, sometimes referred to as the **bell-shaped curve**, or **normal curve**, is very common in many processes. From the life and death process to the process of getting to work on time, the bell-shaped curve is readily apparent in the world of processes. As another example, in looking at heights of adult people, it can be visualized that very few people (with respect to the entire U.S. population) are under a height of 4 ft. tall. Similarly, very few people (with respect to the entire U.S. population) are over 7 ft. tall. Rather, most people's height lie in the 5 ft. to 6 ft. region. The curve for the height of adults will probably look similar to the one shown in Figure 3–4.

As with most normal processes, notice that although the curve for height may not be absolutely perfectly symmetrical, it does take the shape of a bell-shaped curve. It does not take the shape of an exponential curve, binomial curve, or other distribution. The bell-shaped curve can possess different characteristics, such as height and width, and still be classified as a normal curve. Three different bell-shaped curves with different characteristics are shown in Figure 3–5.

The normal curve has two characteristics that are discussed in depth in Section 3.7, Central Tendency, and Section 3.8, Variability. Central tendency is where the curve centers itself, and variability is a degree of consistency. To demonstrate the concept of central tendency and variability before studying them in depth, an example that stresses the normal curve, central tendency, and variability is in order.

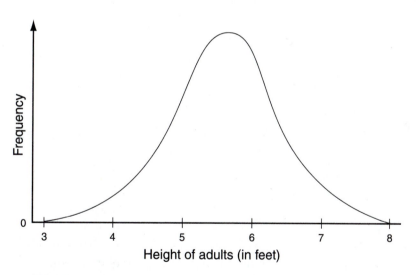

**FIGURE 3–4**
Sample of Normal Curve for Adults' Height

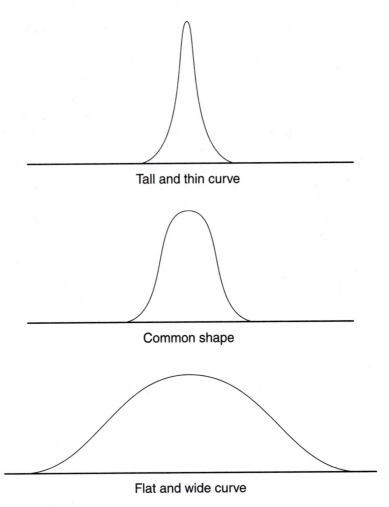

**FIGURE 3–5**
Different Bell-Shaped Curves

EXAMPLE 1   Suppose we are interested in studying the amount of time it routinely takes to get to work. The process can be described as shown:

- Approximately 1 minute to get in the car, start it and leave the house
- An approximately 6-minute drive from the house to the highway
- An approximately 15-minute drive on the highway
- An approximately 6-minute drive from the highway to the parking lot
- An approximately 2-minute walk from the parking lot to the office

It would be extremely elementary to assume that it will take exactly 30 minutes every single day to get to work. There will be variability. Sometimes there will be less traffic on the road, sometimes more. Sometimes the weather will cooperate and sometimes there will be snow or rain or some other weather element that would slow the trip down. Sometimes we may leave earlier in the morning, sometimes later. Notwithstanding, data are gathered for the next thirty days. The data represent the time it takes to get to work as shown in Table 3–1.

**TABLE 3–1**
Results of Time to Get to Work Sample

| Numbers in minutes: | 30 | 31 | 26 | 30 | 27 | 28 |
|---|---|---|---|---|---|---|
| | 29 | 30 | 27 | 28 | 29 | 30 |
| | 35 | 33 | 25 | 30 | 30 | 28 |
| | 32 | 33 | 34 | 30 | 32 | 30 |
| | 29 | 31 | 27 | 29 | 31 | 32 |

The next step after gathering the data is to make a histogram of that data to determine the shape (in this case it is seen in Figure 3–6 that the histogram does not take the shape of a perfectly normal or bell-shaped curve, but does exhibit normal tendencies).

The histogram profiles the process distribution. In our example, it can be seen that the average time it takes to get to work is 29.87 minutes. Likewise, the process possesses

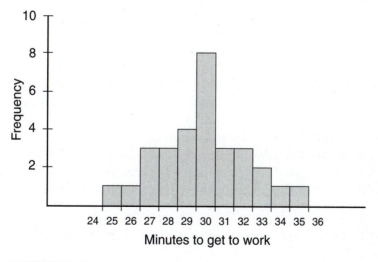

**FIGURE 3–6**
Histogram of Time to Get to Work

variability, containing values from 25 minutes to 35 minutes to get to work. This represents the process for the month that data were collected. It is incorrect to think that the data collected for this month are indicative and representative of the process for the entire year, especially if they were gathered in a location where the climate changes drastically. In this case, the data may only be valid for the month in which they were collected. The bottom line of this example is that the normal curve describes the process of getting to work on time and exhibits both central tendency and variability.

## 3.7 Central Tendency

Central tendency was first addressed in Chapter 2 with respect to accuracy. Central tendency is a mathematical concept that focuses on answering the question "Where do the data tend to center or hover?" There are three main measures of central tendency: **mean**, **median**, and **mode**.

### 3.7.1 Mean

The mean is the most common of the measures of central tendency. The term *mean* refers to the **arithmetic average** of a group of numbers (which is why mean and average are used interchangeably). It is defined as the sum of the values divided by the number of items. The formula for the mean for sample and population is shown as:

| SAMPLE | POPULATION |
|---|---|
| $$\mu_x = \frac{\Sigma x}{n}$$ | $$\mu_x = \frac{\Sigma x}{n}$$ |

where:

$\mu$ = mean of the sample

x = the individual data points

n = the number of items in the sample

where:

$\mu_x$ = the mean of the population

x = the individual data points

n = the number of items in the population

When visualizing the concept of the average, think of the number line as a seesaw with the average being the fulcrum or balancing point. In each of the examples in Figure 3–7, the average is 50. The average not only acts to represent the balancing point, but smooths all of the numbers involved into one number as well.

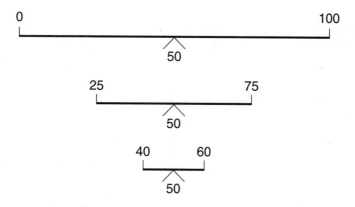

**FIGURE 3–7**
Concept of Average: The Fulcrum of a Seesaw

---

**EXAMPLE 1**   There are three production lines in a manufacturing plant that produce car batteries. A typical day was sampled to determine the average number of batteries produced. The lines produced 500, 547, and 481 batteries on the day that was sampled. The mean number of batteries produced is:

$$\mu = \frac{500 + 547 + 481}{3}$$

$$\mu = 509.3 \text{ batteries}$$

---

## 3.7.2 Median

The median of a group of items is merely the value of the middle item when all of the items in the group are arranged in ascending or descending order. It is the number in a set of data such that one half of the observations are less than that number and one half of the observations are greater than that number. The median is handled differently for groups of data that have an odd number of items than it is for groups of data that have an even number of items. The following two examples show how to compute the median when the number of observations is odd and when it is even.

**EXAMPLE 1**   (Odd number of observations): A manager of the eastern district of department stores is in charge of nine stores. He is interested in finding the median of the number of employees of his stores. He puts the number of employees in ascending order as follows:

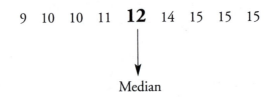

9   10   10   11   **12**   14   15   15   15

Median

The median is twelve employees. There are four observations to the left of 12 and four observations to the right.

**EXAMPLE 2**   (Even number of observations): The manager of the western district needs to determine the median for his twelve stores:

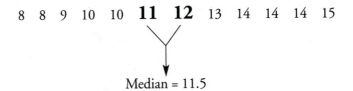

8   8   9   10   10   **11   12**   13   14   14   14   15

Median = 11.5

In this example, it is seen that there exist two numbers in the middle, where there are five to the right and five to the left. The two middle numbers are 11 and 12. Therefore, when the number of observations is even, the average of the two middle numbers is computed to determine the median: (11+12)/2 = 11.5 employees.

### 3.7.3 Mode

By definition, the mode is the most frequently occurring item in a group of data, or even simpler, it is the item that comes up most often.

EXAMPLE 1   A sample of cylinders from a lot is measured for inside diameter. The sample of twenty yielded the measurements shown in Table 3–2.

**TABLE 3–2**
Results of Inside Diameter of Cylinder Sample

| | | | |
|---|---|---|---|
| 45.0 | 46.0 | **45.4** | 45.5 |
| **45.4** | 45.7 | 45.8 | **45.4** |
| **45.4** | 45.1 | 45.6 | 45.2 |
| 45.3 | **45.4** | 45.9 | **45.4** |
| 45.7 | 45.6 | 45.2 | **45.4** |

In this example the mode is 45.4 in. Out of the twenty samples, this measurement (45.4 in.) occurred seven times. It is important to note that while the mode is a measure of central tendency, it may not always be as good of an indicator of central tendency as the mean or median is, as will be shown in Example 2.

EXAMPLE 2   Six trucks are needed for use at a construction site, and tire punctures are of concern. A random sample of ten tires was tested from these trucks and the number of tire punctures per tire was recorded for one week. The number of tire punctures per tire is shown in Table 3–3.

**TABLE 3–3**
Results of Tire Puncture Sample

| | |
|---|---|
| 4 | 3 |
| 6 | 12 |
| 5 | 1 |
| 2 | 7 |
| 8 | 12 |

As can be seen from the data, the mode is 12, which does not give as good an indication as the mean (6.0 punctures) or the median ( 5.5 punctures). The bottom line is, as always, to not become fascinated with the technique as opposed to what the results are saying about the situation.

As a final note in discussing the measures of central tendency, the fact that the mean, median, and mode are all measures of central tendency does not imply that their numeric values will be equal. The only time the mean, median, and mode will be equal is when the distribution is symmetrical. Figure 3–8 demonstrates the relation-

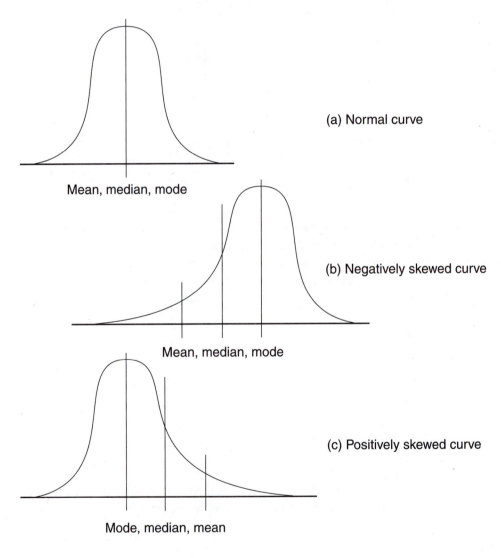

(a) Normal curve

Mean, median, mode

(b) Negatively skewed curve

Mean, median, mode

(c) Positively skewed curve

Mode, median, mean

**FIGURE 3–8**
Relationship of Mean, Median, and Mode to Different Curves

ship of the mean, median, and mode for distributions that are normal, negatively skewed, and positively skewed.

The concept of **bimodality** is worth mentioning before concluding this section. There may be times when there are two modes. This is usually brought about by inadvertently mixing two product streams and then using this mix to determine the mode.

## 3.8 Variability

Similarly to central tendency, variability was first addressed in Chapter 2 as applied to precision. The best way to visualize the variability concept is to recall the discussion of dispersion as it applied to the archery, where the idea is to hit the bullseye as consistently as possible, as shown in Figure 3–9.

Figure 3–9 depicts minimal variability or dispersion. Unfortunately, the situation depicted in Figure 3–9 may not always be realistic and the shots to the target may contain variability, or dispersion, as shown in Figure 3–10.

The two most common terms to describe variation used in quality are the **range** and the **standard deviation**.

### 3.8.1 Range

The range is the difference between the largest and smallest values in the group. Because the range is a measure of variation, if there is no difference between the largest and smallest numbers, there is no variation. Stated another way, the numbers within the group are very consistent.

**FIGURE 3–9**
Bullseye with Minimum Variation

**FIGURE 3–10**
Bullseye with Large Variation

---

**EXAMPLE 1**   Two different manufacturing lines produce surface plates. Eight surface plates from each line are tested for hardness. The hardness numbers for the eight surface plates from each line and the associated range for each are as follows:

LINE 1:

100

99

98

100   RANGE: 100 (highest value) – 98 (lowest value) = 2

100

100

99

98

LINE 2:

102

99

100

100   RANGE: 102 (highest value) – 90 (lowest value) = 12

104

98

100

90

---

When considering the range, the larger the range, the larger the variation. In our example, a numerical value of 2 is much smaller than a numerical value of 12, thus the variation of the surface plates produced by Line 1 is smaller than the variation of those produced by Line 2. Notice that we are not addressing statistical outliers at this point. Although there definitely is a statistical theorem that addresses outliers, in the world of quality these outliers are not thrown out prematurely, because there may be much information obtained from them.

## 3.8.2 Standard Deviation

The standard deviation is used when the data are normal, or bell-shaped. Although this may be viewed as a restriction, a large number of the processes and data that measure physical properties such as temperature, length, and mass will exhibit normal tendencies, as was discussed earlier. The standard deviation can be computed for the population or for the sample, with the only difference being in the formula: the formula for a population does not include a degree of freedom in the denominator, whereas the standard deviation for a sample does:

SAMPLE                                    POPULATION

$$s_x = \sqrt{\frac{\Sigma(X - Xbar)^2}{n-1}} \qquad\qquad \sigma_x = \sqrt{\frac{\Sigma(X - \mu_x)^2}{N}}$$

where:                                    where:

$s_x$ = the sample standard deviation     $\sigma_x$ = the population standard deviation

X = the individual sample data points     X = the individual data points

Xbar = the mean of the sample             $\mu_x$ = the number of items in the population

n = the sample size                       N = the population size

The numerical value of the standard deviation gives insight into the spread of the distribution. The smaller the numerical value of the standard deviation, the smaller (or

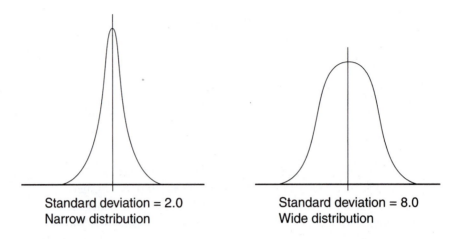

Standard deviation = 2.0
Narrow distribution

Standard deviation = 8.0
Wide distribution

**FIGURE 3–11**
Relationship of Standard Deviation to Normal Curve Shape

narrower) the spread of the distribution. The larger the numerical value for standard deviation, the larger the variation of the distribution. For example, the two distributions in Figure 3–11 have a standard deviation value of 2.0 and 8.0, respectively.

In comparing the two distributions, it can be seen that in addition to the standard deviation giving insight into the variability, there is more information that the standard deviation tells with respect to how the process is performing.

---

**EXAMPLE 1**   Return to Example 1 in Section 3.6, which pertains to the time it routinely takes to get to work. Recall that a sample of one month out of the year was collected. By using the formula, the sample standard deviation is computed to be 2.33 minutes. This gives us insight into where this process wants to operate most of the time:

- The area represented by the mean ± 1 standard deviations is the area where the process resides 68.26% of the time. The 68.26% is an approximation and is found by taking the area of the curve. This approximation is good for all normal curves, regardless of their shape or height.
- The area represented by the mean ± 2 standard deviations is the area where the process resides 95.45% of the time.
- The area represented by the mean ± 3 standard deviations is the area where the process resides 99.728% of the time.

This is pictorially represented in Figure 3–12.

The standard deviation associated with the process can explain in layman's terms what is happening with the process of getting to work:

- The process averages 29.87 minutes with a standard deviation of 2.33 minutes.
- One could expect to get to work between 27.54 minutes and 32.20 minutes (mean ± 1 standard deviation) 68.26% of the time
- One could expect to get to work between 25.21 minutes and 34.53 minutes (mean ± 2 standard deviations) 95.45% of the time
- One could expect to get to work between 22.88 minutes and 36.86 minutes (mean ± 3 standard deviations) 99.728% of the time

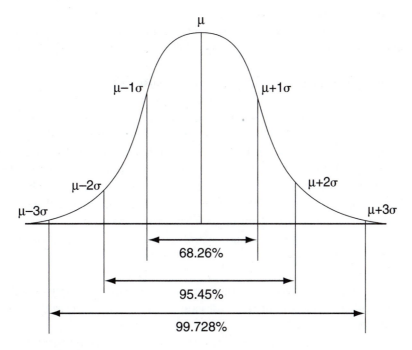

**FIGURE 3–12**
Area Under the Curve with Respect to Standard Deviations

- NOTE: the normal curve is a continuum and goes out in both directions indefinitely, and although ± 3 standard deviations is most commonly the focus of attention, it is important to note that:
  - >> There is 99.9937% of the area under the curve for the mean ± 4 standard deviations.
  - >> There is 99.999943% of the area under the curve for the mean ± 5 standard deviations.
  - >> There is 99.9999998% of the area under the curve for the mean ± 6 standard deviations.

In the example the standard deviation gave insight into the variability of the process and the boundaries for where the process operates given the way one goes to work, the driving habits, etc. Furthermore, as will be seen in Chapter 8, the mean ± 3 standard deviations has significant meaning in control charting, which is why an understanding of standard deviation is important.

As a final point to make in this example, the probability that one would get to work in less than 22.88 minutes is very small. Likewise, the probability that one would take longer than 36.86 minutes to get to work is very small as well. So, what would the probability be that someone could expect to fall outside of these specific values, which also constitute the mean ± 3 standard deviations? Approximately .072% of the time (100% − 99.728%). When the data points fall outside of the mean ± 3 standard deviations, it is indicative of "special cause" variation (special cause variation is discussed in detail in Chapter 8). As an example of special cause variation, there may be an automobile accident on the highway that prevents someone from getting to work within the boundaries of 22.88 and 36.86 minutes. Notice that these special causes do not happen every day. They happen occasionally. This concept is extremely important and as stated above, will be further exploited as progression to Statistical Process Control (SPC) occurs.

Finally, it is vitally important to realize some additional key concepts associated with standard deviations:

- The mean ± 1 standard deviation accounts for approximately 68.26%, or 2/3, of the process. This boundary is sometimes referred to as where the process operates "most of the time."
- The data from the process determines the value of the standard deviation. No member of management, worker, or anyone else for that matter, can dictate that a process will exhibit a particular amount of variation. It is the process and the process data that determine the standard deviation. However, because management determines the manpower, methods, materials, and machinery (this is discussed in Chapter 4), management does make the process changes that lead to variability reduction.

- The range (highest minus lowest values) of a normal distribution usually approximates the mean ± 3 standard deviations, but it is not an exact approximation.

---

**EXAMPLE 2**   There is concern over the number of defects that are coming off a cam production line. A month of data were collected that revealed that the average number of errors (an error is defined as one critical dimension on the cam that is out of tolerance) was 6 with a standard deviation of 1. From these data, the following information is determined:

- Given the way the production line operates (the manpower, methods, materials, and machinery), between five and seven defects can be expected to occur 68.26% of the time.
- Given the production line process, between four and eight defects can be expected to occur 95.45% of the time.
- Given the production line process, between three and nine defects can be expected to occur 99.728% of the time.

---

# 3.9 Areas Under the Normal Curve

There are times in the world of quality when the amount of scrap and rework needs to be calculated. Likewise, there are times when it is desirable to calculate certain areas under the normal curve. It is in these situations that the **Z-test** is invaluable and is the statistical tool used to compute these areas or probability of occurrence.

## 3.9.1 The Z-Statistic

The Z-test is one of the mathematical tests used to compute the area under the curve. More specifically, the Z-test computes the area to the left of some particular point of interest. To conduct the Z-test, a Z-statistic is first computed based on the point of interest, the mean, and the standard deviation. The Z-statistic value is then looked up in a Z-table. The Z-table is the only left reading table in statistics and depicts the area under the normal curve to the left of that point of interest. Should the area to the right of a point of interest be needed, then the following procedure would be used:

- Calculate the area to the left of the point of interest using the Z-table.
- Subtract the area to the left from 1 (recall from statistics that the area under a normal curve is equal to 1).

**TABLE 3–4**

Results of Tire Life Test Sample

| Data in thousands of miles: | 75 | 75 | 80 |
|---|---|---|---|
| | 70 | 75 | 75 |
| | 77 | 73 | 77 |
| | 73 | 72 | 79 |

The best way to explain the theory associated with the Z-statistic is to start with an example. In Table 3–4 a sample of tires is life tested with the raw data indicated.

The histogram for the tires is shown in Figure 3–13.

The curve is normal with a mean of 75.08 (or 75,080 miles) and a standard deviation of 2.87 (or 2870 miles). Suppose that the quality department is interested in seeing what percentage of the population of tires lasts more than 79,000 miles, as shown in Figure 3–14.

To do this, the Z-statistic formula is used and the numbers from the example are plugged into the equation:

$$\text{Formula for Z:} \quad Z = \frac{x - \mu}{\sigma}$$

$$\textit{where:} \quad X = \text{the data point of interest}$$
$$\mu = \text{the process mean}$$
$$\sigma = \text{the process standard deviation}$$

$$\text{substituting:} \quad Z = \frac{79 - 75.08}{2.87}$$
$$Z = 1.37$$

After determining the Z value, the Z-table in the appendix will determine the area to the left of the point of interest (79,000 miles). The area to the left of 79,000 miles is .9147 or 91.47%. Interpreted this means that 91.47% of the tires produced will last less than 79,000 miles. To find the percentage of tires that will last longer than 79,000 miles:

$$1 - .9147 = .0853 = 8.53\%$$

The pictorial representation is shown in Figure 3–15.

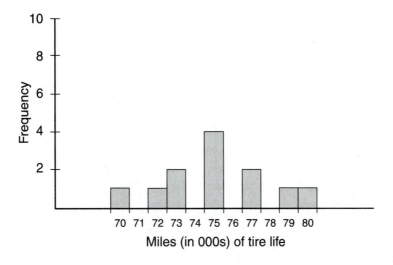

**FIGURE 3–13**
Tire Life Histogram

There are likewise times when it is necessary to know the percentage of time a product will perform between two values that lie within the distribution, and the next example demonstrates how the Z-test would be used to handle this type of scenario.

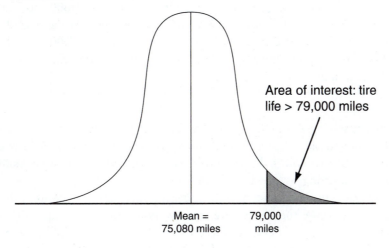

**FIGURE 3–14**
Area Under the Curve: Tire Life Greater than 79,000 Miles

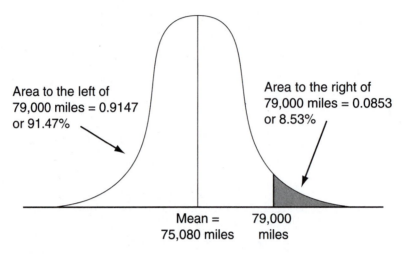

Area to the left of
79,000 miles = 0.9147
or 91.47%

Area to the right of
79,000 miles = 0.0853
or 8.53%

Mean =     79,000
75,080 miles   miles

**FIGURE 3–15**
Area Under the Curve: Tire Life Less than 79,000 Miles
and Greater than 79,000 Miles

**EXAMPLE 1**    In continuing the example, suppose that the quality department is now interested in knowing what percentage of tire production lies between 71,000 and 72,000 miles as shown in Figure 3–16. From the diagram it can be seen that the shaded area is what needs to be determined. This is the area between Point 1 (71,000 miles) and Point 2 (72,000 miles), and the heavy arrows indicate the area that the Z-test will yield for each point of interest.

Before solving this problem with the Z-table, it is time to review a little bit about subtracting areas. When two areas are subtracted, what will remain is the distance between the two points. To prove this very simply, take a 1 in. string and subtract it from a 3-in. string, as shown in Figure 3–17.

This same theory applies to areas under the curve as well: one area subtracted from another area leaves the area between the two points. In our example, the area of interest is the percentage of tires produced that will have a tire life of between 71,000 and 72,000 miles. The area to the left of 72,000 miles will be computed, the area to the left of 71,000 miles will be computed, and the two areas are then subtracted. Note that the areas must be subtracted because you can never subtract Z-values from one another and then look up the difference in the table for the area!

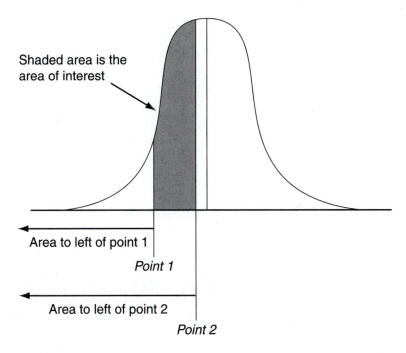

**FIGURE 3–16**
Area Under the Curve: Greater than 71,000 Miles and Less than 72,000 Miles

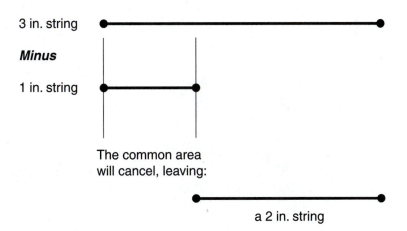

**FIGURE 3–17**
Concept of Subtracting String Lengths

$$Z = \frac{X - \mu}{\sigma}$$

for 71,000 miles:   $Z = \dfrac{71 - 75.08}{2.87}$

$Z = -1.42$

$Z_{-1.42} = 0.0778$

for 72,000 miles:   $Z = \dfrac{72 - 75.08}{2.87}$

$Z = -1.07$

$Z_{-1.07} = 0.1423$

area between 71,000 and 72,000 miles:

$0.1423 - 0.0778 = 0.065$ or 6.5%

The pictorial for this situation is shown in Figure 3–18.

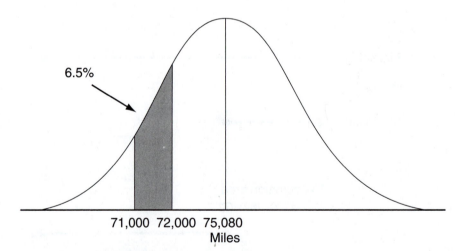

**FIGURE 3–18**
Area Under the Curve for Tire Life Greater than 71,000 Miles
and Less than 72,000 Miles

## 3.9.2 An Excel Spreadsheet for Computing the Z-Statistic and Area Under the Curve

An Excel spreadsheet is provided in Figure 3–19, and in the CD at the back of the text, for computing the area under the curve with respect to a point of interest. The program is shown in Figure 3–19 using the data from Example 1 in Section 3.9.1 (relating tires that lasted more than 79,000 miles).

The details of the spreadsheet are as follows:

- The shaded cells represent data entry in column B of the spreadsheet: the mean of the process, the standard deviation of the process, and the point of interest.
- The Z-statistic is computed via the formula (point of interest – mean) divided by standard deviation.
- The area to the left of the point of interest is calculated using the NORMSDIST function of the Z-statistic and expressing the value as a percentage.
- The area to the right of the point of interest is calculated by taking 1 minus the NORMSDIST function of the Z-statistic and expressing it as a percentage.

NOTE: The values from the Excel spreadsheet may differ slightly from those computed by hand. The reason is that we round off when computing by hand for using the Z-table, but Excel will not round off before using NORMSDIST:

Z-statistic computed by hand and rounding off: 1.37

Z-statistic computed and used by Excel: 1.3658537 . . .

| | | |
|---|---|---|
| Process mean | 75.08 | |
| Process standard deviation | 2.87 | |
| Point of interest | 79 | |
| Z-statistic | 1.366 | |
| Area to the left of the point of interest | 91.40% | |
| Area to the right of the point of interest | 8.60% | |

**FIGURE 3–19**
Excel Spreadsheet for Z-Statistic

Even with the difference in rounding, it can be seen that this difference is really negligible when using the Z-statistic in real life. The bottom line on the tire example is that about 91.4% or 91.5% of the tires lasted 79,000 miles or less. One would seriously doubt that the quality department personnel using the information would split hairs over whether it was 91.4% or 91.5%.

## 3.10  Other Statistical Concepts

Although a total course could be devoted to just statistics, some remaining concepts will be covered that have a direct relationship to quality and the material covered in this text. Those concepts are basic **probability**, conditional probability, and the concept of attribute data and goalpost mentality.

### 3.10.1 Simple Probability

Suppose that some operation results in one of a definite set of outcomes, but associated with the outcome is randomness (or chance). In other words, the outcome cannot be predicted with certainty. Examples of this type of situation include rolling two dice, drawing a card at random from a deck of shuffled cards, picking the winning raffle ticket from a barrel of tickets, and selecting a part from an incoming lot of items from the production line for a nondestructive performance test.

Assume that some particular outcome is desired: a 1 and 3 when rolling two dice, drawing the queen of hearts from a deck of shuffled cards, picking the ticket with the name Smith, and selecting a part that will pass the nondestructive performance test. With the above stated conditions the definition of probability can be defined as:

$$\text{Probability} = \frac{\text{Number of favorable outcomes}}{\text{Total number of outcomes}}$$

Probability values range includes from 0 (the event has no chance of occurring, or the event has not occurred) up to and including 1 (the event has already occurred). For example, if there is a 100% chance of rain, it is safe to assume that it is raining.

---

EXAMPLE 1    At the annual St. Mark's festival, a game at one of the booths consists of wagering $1 on a number between 1 and 6 and then rolling a single die. If the number on the die is what was wagered, $2 is won. The probability of winning $2 is 1/6 (one favorable outcome and six total possible outcomes).

---

**EXAMPLE 2**   The next booth at the St. Mark's festival is a game of cards. A wager of $1 will yield a $5 return if the right card value is bet (for example, it is not necessary to select the value AND suit, just the card value). The probability of winning is 4/52 because there are 4 possibilities (for example, four kings, four aces, four jacks, or four of any card value) out of 52 possible outcomes.

## 3.10.2 Addition Principle of Probability

If one group contains "x" items, another group contains "y" items, and "z" represents the items that are in both groups, then the number of items that are in both two groups is x + y – z. Likewise, when dealing with probabilities, the equation becomes (where E1 is the first event and E2 is the second event):

$$Pr\{E1 \text{ or } E2\} = Pr\{E1\} + Pr\{E2\} - Pr\{E1 \text{ and } E2\}$$

**EXAMPLE 1**   In the Department of Mathematics and Statistics at a local college, one slice of the student population is broken down as depicted in Table 3–5.

**TABLE 3–5**
Department of Mathematics and Statistics Student Breakdown (Raw Count)

|        | MATHEMATICS | STATISTICS | TOTAL |
|--------|-------------|------------|-------|
| Male   | 40          | 50         | *90*  |
| Female | 20          | 55         | *75*  |
| **TOTAL** | *60*     | *105*      | *165* |

- the number of students who are either female OR in statistics equals:

  75(total females) + 105(total statistics) – 55(female and statistics) = 125

- the number of students who are either male OR in mathematics equals:

  90(total males) + 60(total mathematics) – 40(male and mathematics) = 110

**EXAMPLE 2**   The raw counts were used in EXAMPLE 1. In this example, the raw counts are replaced with the respective probabilities, which are determined by taking the number in the particular category and dividing by the total number. For example, for the number of males in mathematics: 40/165 = .242. The results are shown in Table 3–6.

**TABLE 3–6**
Department of Mathematics and Statistics Student Breakdown (Probability)

|        | MATHEMATICS | STATISTICS | TOTAL |
|--------|-------------|------------|-------|
| Male   | .242        | .303       | .545  |
| Female | .121        | .333       | .454  |
| **TOTAL** | .363     | .636       | .999  |

- The probability that a student will be either female OR in statistics equals:

  .454(total females) + .636(total statistics) − .333(female and statistics) = .757

  And in checking against the results obtained in Part A of EXAMPLE 1:

  $$?$$

  (Total number of students)(Pr(female or statistics)) = 125

  (165)                     (0.757) = 124.9 = 125   The answer is correct.

- The number of students who are either male OR in mathematics equals:

  .545(total males) + .363(total mathematics) − .242(male and mathematics) = .666

  And in checking against the results obtained in Part B of Example 1:

  $$?$$

  (Total number of students)(Pr(male or mathematics)) = 110

  (165)                     (0.666) = 109.9 = 110 The answer is correct.

The rules of probability apply to quality issues as well. For example, a random sample of tires could be obtained from a production line and road tested under various conditions (for example, highway driving and city driving) to answer questions such as:

- What is the probability that a tire tested under city driving conditions experiences tire wear in excess of 0.1" OR experiences two tire punctures?
- What is the probability that a tire tested under highway conditions experiences tire wear in excess of 0.1" OR experiences two tire punctures?

### 3.10.3 Conditional Probability

Frequently, there exists a need to determine the probability of an event occurring given the fact that some other event has already occurred. This type of probability is called conditional probability and is expressed as Pr{A/B}, where A and B are different events and event B has already occurred. In words it is expressed as "the probability that event A will occur GIVEN the fact that event B has already occurred."

Tires can go bad, regardless of whether a warranty has been purchased or not. Likewise, tires may not go bad, regardless of whether or not a warranty has been purchased. The warranty department of a northern tire company is interested in gathering some information on the quality of their company's tires so that a fair warranty policy can be determined. They are further interested in the number of tires that go bad when a warranty has been purchased, the number of tires that remain good when a warranty has been purchased, the number of tires that go bad when no warranty has been purchased, and the number of tires that remain good when no warranty has been purchased. Data are collected over a period of six months for the tire warranty department and are portrayed in Table 3–7.

- What is the probability that a selected tire will be good given the fact that a warranty has been purchased? In statistical terms the probability equation is Pr{good/warranty}. The first thing to do is to look at the denominator (warranty). A warranty has already been purchased so the denominator is 500. The next step is to look at

**TABLE 3–7**
Conditional Probability Table for Warranty and Non-Warranty Tires

|  | WARRANTY | NO WARRANTY | TOTAL |
|---|---|---|---|
| Good | 200 | 400 | 600 |
| Bad | 300 | 200 | 500 |
| **TOTAL** | 500 | 600 | 1100 |

the numerator (a selected tire is good) that is in the column under warranty (because warranty has already occurred). The number is 200, thus the probability is calculated:

Pr{good tire/warranty} = 200/500 = .4 or 40%

The same philosophy is used for the remaining answers for the warranty department.

• What is the probability that a selected tire is bad given the fact that a warranty has been purchased? The event of a warranty being purchased has occurred, so the denominator is 500. The numerator (a tire selected is bad) is 300. Thus,

Pr{bad tire/warranty} = 300/500 = .6 or 60%

• What is the probability that a selected tire will be good given the fact that no warranty has been purchased?

Pr{good tire/no warranty} = 400/600 = .667 = 66.7%

• What is the probability that a selected tire will go bad given the fact that no warranty has been purchased?

Pr{bad tire/no warranty} = 200/600 = .333 = 33.3%

Notice that when the conditional probability is computed for each given event (warranty and no warranty) that the sum of the two unknown outcomes (good tires and bad tires) will always equal 1 or 100%:

$$
\begin{array}{ccccc}
 & & & & ? \\
\text{Pr\{good tire/warranty\}} & + & \text{Pr\{bad tire/no warranty\}} & = & 1 \\
.4 & + & .6 & = & 1.0
\end{array}
$$

$$
\begin{array}{ccccc}
 & & & & ? \\
\text{Pr\{good tire/no warranty\}} & + & \text{Pr\{bad tire/no warranty\}} & = & 1 \\
.667 & + & .333 & = & 1.0
\end{array}
$$

The concepts of probability are used in the world of quality. Insight into product performance and warranties are areas where probability can be used.

### 3.10.4 The Concept of Attribute Data and "Goalpost" Thinking (aka Tolerance Stack-Up)

Although attribute data are a legitimate type of data, caution must be exercised in their intended usage (as with any type of data). The danger with a pass/fail type of classification is apparent when dealing with specifications. Most specifications have a tolerance with them (for example, the steel rod must be 5 in. in length ± 0.1 in.). Although the intent of the tolerance is necessary (because no two items are exactly identical), some complacency has occurred in that many people have begun to believe that "all parts that lie within the tolerance" will perform in the same manner. Although this mentality may be viable in football, where any field goal between the goalposts is good, it is not viable in manufacturing. Even though the parts fall in tolerance, there will be some variability in their performance. In the case of the steel rods, two rods that measure 4.9 in. and 5.1 in. will both be accepted as "good" or "in-tolerance." However, it is not necessarily correct to assume that both will function exactly the same in all aspects. The attribute thinking associated with tolerances is shown in Figure 3–20.

The thinking that "anything that lies within the tolerance is acceptable" is sometimes prevalent in manufacturing and other company organizations. This thought is definitely not optimal, especially when assembling multiple components with multiple tolerances. This is when **tolerance stack-up** may occur as well as increased product and product performance variability. Figure 3–21 depicts what occurs when two parts are mated together (each with different but acceptable tolerances):

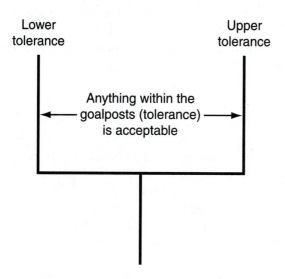

**FIGURE 3–20**
Goalpost Concept with Upper and Lower Tolerances

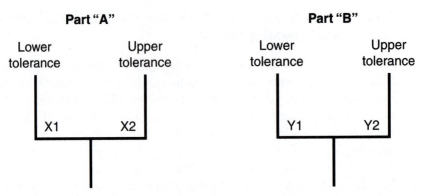

Part X1 when mated with Part Y1 will not perform *exactly* the same as Part X2 that is mated with Part Y2

**FIGURE 3–21**
Goalpost Concept When Mating Parts

In addition to performance variability between parts X1Y1 and X2Y2, the overall tolerance of the two mated parts is different and may or may not be in spec for the assembled component. This is where tolerance stack-up (where the tolerances of the individual components add up to create an out of tolerance assembly) may occur. Although it is not the intent of this text to go into the depth of tolerances (that is an entire course in itself), it is its intent to make the reader aware of this type of thinking as applied to the world of quality. The main point is that when using attribute data (or any data for that matter), understanding of the data collected from the process and data implications is paramount.

## 3.11 Summary

- When it is determined that a sample needs to be drawn from a population, it is imperative that the sample is drawn randomly. If it is not drawn randomly then the confidence level and margin of error associated with the sample will be adversely affected. This, in turn, will negatively affect the accuracy of the decisions made with respect to the processes that generate products and services.
- A quick and rough way for approximating the margin of error associated with a sample is to divide 1 by the square root of the sample size.
- Surveys are an excellent instrument for collecting data that may not already exist or be readily available, but care must be taken to ensure that the survey questions are not biased or leading. In addition, it is imperative to ensure that the survey is dis-

tributed randomly among the target population. Otherwise, the results will not be correct due to the effects of bias.

- The steps of a survey are:

  1. Identify the problem area or area of focus for the survey.
  2. Conduct a thorough literary search.
  3. Predevelop the stage of the survey.
  4. Develop questions using a survey checklist.

- The normal, or bell-shaped curve, is the most common mathematical distribution.
- The normal curve is characterized by central tendency (using the mathematical measures of mean, median, and mode) and variability (using the mathematical measures of range, standard deviation).
- When combined with the mean, the standard deviation gives insight into where the process operates:

  ➢ 68.26% of the time the process will operate in the area of the mean ± 1 standard deviation.
  ➢ 95.45% of the time the process will operate in the area of the mean ± 2 standard deviations.
  ➢ 99.728% of the time the process will operate in the area of the mean ± 3 standard deviations.

- The normal curve is the fundamental theory behind the statistical process control (SPC) chart, which will be explored in detail in Chapter 8.
- The goalpost mentality occurs when parts are accepted as long as they lie within the lower specification limit (LSL) and upper specification limit (USL). This is similar to kicking a field goal in football: the attempt is good as long as the football goes between the goalposts (regardless of where it goes through). Care must be taken when using the goalpost theory because it is a major contributor to tolerance stack-up.
- The Z-statistic is a technique used to evaluate the area under a curve from the point of interest leftward. Likewise, the Z-statistic can be used to find the area to the right of the point of interest (1 minus the area to the left of the point) and between two points of interest as well.
- Probability is defined as the number of favorable outcomes divided by the total number of outcomes. Probability values range from 0 (the event has not occurred, or the event has no chance of occurring) to 1 (the event has already occurred).
- Conditional probability has to do with determining the probability of an event occurring, given the fact that some other event has already occurred.

# Review Questions

1. The margin of error of a sample can be estimated by:
   a. using the Z-table
   b. dividing 1 by the square root of the sample size
   c. matching to the appropriate confidence levels
   d. dividing 1 by the sample size

2. The standard deviation:
   a. gives an approximation of the size of the central tendency
   b. uses a degree of freedom in the formula for a population
   c. gives an idea of the consistency of the process
   d. is a measure of central tendency

3. Conditional probability is based on the premise that:
   a. one event has not occurred while the other event has already occurred
   b. neither events have occurred
   c. both events have occurred
   d. none of the above

4. In a perfectly normal curve:
   a. the mean and mode will always be equal
   b. the mean and the median will always be equal
   c. the mean, median, and mode will always be equal
   d. the median and mode will always be equal

5. Approximately two thirds of the data in a normal curve:
   a. are included in the mean ± 1 standard deviation
   b. are included in the mean ± 2 standard deviations
   c. are included in the mean ± 3 standard deviations
   d. none of the above

6. The probability associated with an event:
   a. will lie between 0 and 1
   b. will be very close to 0 if the event is far off in the future
   c. will be very close to 1 if the event is far off in the future
   d. equals the number of favorable outcomes divided by the total number of outcomes

7. Which of the following is not an example of random number selection for a sample of eighteen items?
   a. Using a random number table
   b. Using a computer random number generator

   c. Rolling three dice
   d. All of the above

8. If the mean, median, and mode are all different, one can conclude:
   a. that the distribution is normal
   b. that the distribution may contain skewing
   c. that the distribution is binomial
   d. nothing can be concluded

9. A good technique for obtaining data that may not be readily available is:
   a. cold calls to hand-selected individuals
   b. surveys to hand-selected individuals
   c. estimations about the population using measures of central tendency and variability
   d. surveys to randomly selected individuals

10. Approximately one third of the data in a population that are normally distributed:
   a. occur when the range is 0
   b. are included in the mean ± 1 standard deviation
   c. occur when the range is greater than 1
   d. none of the above

11. Mapping out the parts of a survey, generating hypothesis about the data being collected, and determining the sample size, confidence level, and margin of error are all part of:
   a. identification of the problem area or area of focus
   b. the literary research
   c. the predevelopment stage
   d. none of the above

12. Distribution A has a standard deviation of 4 while distribution B has a standard deviation of 6. What can be concluded?
   a. A has greater variability than B.
   b. B has greater variability than A.
   c. The average of each distribution is needed to answer the question correctly.
   d. None of the above.

13. The main contributor to tolerance stack-up is:
   a. lack of process control
   b. the range of a process being greater than 0
   c. a standard deviation that is equal to the mean
   d. reliance on the goalpost mentality
   e. none of the above

14. When developing survey questions, it is important to include all but:
    a. ensuring that the sample group is demographically homogeneous
    b. ensuring that the questions could not be misinterpreted
    c. ensuring that the questions do not include grammar that is very unfamiliar
    d. assuming that assumptions do not exist in the questions
    e. none of the above

15. How much of the data in a normal curve are contained in the area of the mean ± 3 standard deviations?
    a. 68.26%
    b. 95.45%
    c. 99.728%
    d. Upper control limit
    e. None of the above

16. Which technique is best for determining the area under the normal curve?
    a. Z-statistic
    b. Central tendency
    c. Variability
    d. None of the above

17. In a positively skewed distribution:
    a. the mode is to the left of the median
    b. the mean is to the left of the mode
    c. the mean is equal to the mode
    d. all of the above
    e. none of the above

18. Determining the probability of either event 1 or event 2 occurring is an example of:
    a. conditional probability
    b. the addition principle of probability
    c. simple probability
    d. none of the above

19. Types of survey questions should not include:
    a. multiple choice
    b. open ended
    c. two-way
    d. all of the above

20. The most common occurring value in a normally distributed curve is:
    a. the mode
    b. the average

    c. the median

    d. the range

    e. none of the above

21. In a negatively skewed distribution:
    a. the mode is to the left of the median
    b. the mean is to the left of the mode
    c. the mean is equal to the mode
    d. all of the above
    e. none of the above

22. The area under the normal curve in the region of the mean ± 4 standard deviations is:
    a. 99.728%
    b. 99.9937%
    c. 99.9999%
    d. none of the above

23. Approximately 95% of the data in a population that are normally distributed:
    a. occurs when the range is 0
    b. are included within the area of the mean ± 1 standard deviation
    c. are included within the area of the mean ± 2 standard deviations
    d. occur when the probability is 0.95
    e. none of the above

24. When conducting something other than a 100% sample, the following condition will occur:
    a. margin of error
    b. standard deviation
    c. central tendency
    d. all of the above
    e. none of the above

25. Distribution A has a range of 10 while Distribution B has a range of 15. What can be concluded?
    a. A has a greater variability than B.
    b. B has a greater variability than A.
    c. The range is equal to the mean.
    d. The average of each distribution is needed to answer the question correctly.
    e. None of the above.

26. Distribution A has an average of 5 while Distribution B has an average of 7. What can be concluded?
    a. A will be a better process for giving the customer a quality product or service.

    b. B will be a better process for giving the customer a quality product or service.

    c. A measure of variability is needed to answer the question correctly.

    d. None of the above.

27. The Z-table:
    a. is a left reading table in mathematics
    b. determines the area under the normal curve
    c. only has values from −3.50 to +3.50, inclusive
    d. a and b
    e. none of the above

28. To compute a standard deviation for a population:
    a. a degree of freedom in the formula is not necessary
    b. a degree of freedom in the formula is necessary
    c. computation of the variance first is recommended
    d. none of the above

29. To measure some characteristic whose results will be indicative of the entire population is the goal of:
    a. the standard deviation
    b. a survey
    c. a sample
    d. none of the above

30. When conducting a survey, it is highly desirable to have:
    a. a confidence level no lower than 98%
    b. a margin of error as close to 0 as possible
    c. a sample size of at least 10% of the population size
    d. none of the above

## Problems

1. A fishing lure company produces an average of five handmade lures per hour with a standard deviation of one lure per hour. The process has been studied and has been determined to be normal. What percent of the time will the company produce more than six lures per hour?

2. In problem 1, what is the percent of time the company will produce between three and five lures per hour?

3. Raw production data are collected from a company that produces dead bolt locks for residential doors. For the past two years, monthly quantities of a one-time production run of a particular dead bolt are recorded as shown in Table 3–8.

**TABLE 3–8**
Dead Bolt Lock Data

| | | | |
|---|---|---|---|
| 14800 | 16000 | 15700 | 16800 |
| 14200 | 15200 | 15100 | 15900 |
| 14800 | 17000 | 13600 | 16500 |
| 17500 | 16200 | 15200 | 15300 |
| 15200 | 14800 | 14500 | 14300 |
| 13800 | 15400 | 14300 | 15200 |

a. What is the average monthly production?
b. What is the median monthly production?
c. What is the mode monthly production?
d. How do the measures of central tendency compare? What can be said about the distribution with respect to central tendency?

4. For the data in problem 3:
   a. What is the range?
   b. What is the standard deviation?

5. Compute for problem 3:
   a. the mean ± 1 standard deviation
   b. the mean ± 2 standard deviations
   c. the mean ± 3 standard deviations

6. Draw the histogram for problem 3, labeling the mean, mean ± 1 standard deviation, mean ± 2 standard deviations, and mean ± 3 standard deviations.

7. What percent of time can the company be expected to produce between 15,500 and 16,800 dead bolt locks per month?

8. A warranty department is studying the results of warranty and service claims over the first month's sales of their newly introduced mechanical wristwatch and their quartz wristwatch. Answer the questions for the following compiled data shown in Table 3–9.

**TABLE 3–9**
Conditional Probability Table for Wristwatches

|  | PROBLEMS REPORTED (PR) | NO PROBLEMS REPORTED (NPR) |
|---|---|---|
| Mechanical (M) Wristwatch | 25 | 14 |
| Quartz (Q) Wristwatch | 20 | 13 |

a. Pr(PR)
b. Pr(NPR)
c. Pr(PR/M)
d. Pr(NPR/M)
e. Pr(M)
f. Pr(Q)
g. Pr(PR/Q)
h. Pr(NPR/Q)
i. What is the probability that a watch will either be quartz or experience problems?
j. What is the probability that a watch will either be quartz or not experience problems?
k. What is the probability that a watch will be either mechanical or experience problems?
l. What is the probability that a watch will be either mechanical or not experience problems?

9. Answer the following true or false question and explain the reasoning behind your answer:
   "A standard deviation can have a value of 0." True or False

10. Compute the standard deviation for a one-time only production run for a long-time customer consisting of twenty sampled items with the data shown in Table 3–10.

**TABLE 3–10**
Data for Standard Deviation

| | | | |
|---|---|---|---|
| 100 | 104 | 99 | 101 |
| 102 | 99 | 100 | 101 |
| 98 | 99 | 103 | 102 |
| 101 | 97 | 98 | 96 |
| 100 | 100 | 96 | 103 |

11. In Problem 10:
    a. What is the value for the average – 30?
    b. What is the value for the average + 30?

12. Assuming problem 10 was now set up as a continuous production run and the data are representative of a typical lot run, what percent of the population would be:
    a. greater than or equal to 100
    b. less than or equal to 99

13. Determine the probability for each of the following.
    a. An even number occurs in one toss of a fair die.
    b. The sum of 6 occurs in a single roll of two dice.

14. Starting with row 10 in column 2 of the random number table in the appendix, determine which five items will be drawn from a population of 100.

15. Starting with row 15 in column 4 of the random number table in the appendix, determine which five items will be drawn from the same population of 100 in problem 14.

# 4

# Processes

## 4.1 Chapter Objectives

After completing this chapter, the student should be able to:

- Understand the concept of processes.
- Gain an appreciation for the importance of a detailed understanding of processes before implementing improvements.
- Be able to apply the Process Checklist to processes.
- Understand other concepts associated with processes such as process tasks and process measurement.

## 4.2 Key Terms

| | | |
|---|---|---|
| Flowchart | Process | Tweaked |
| *M*s of a Process | Process Checklist | |

## 4.3 Introduction

The final fundamental concept that needs to be addressed is that of **processes**. The initial perception of processes may be that they are not that complicated; that they can be managed with relative ease; and that arbitrary changes can be made to the process to make improvements. As will be seen in this chapter, that is not the case. The concept of processes is really the focus of the first three chapters. It is about processes that the fundamental concepts of quality apply. It is processes that are measured. It is process measurements that are the focus of statistical techniques and applications. This chapter dives into the depths of processes and sets the stage for the roadmap to quality.

## 4.4  Processes: What Are They?

A process is a set of usually sequential value-added tasks that use organizational resources to produce a product or service. A process can be unique to a single department or can cross many departments within an organization. A process is usually repetitive in nature and is needed to make a product or achieve a result. The quality of the product or service is totally dependent on the quality of the process. A process consists of resources that are called the **Ms of a process**:

- Manpower: this represents the personnel associated with making the process operate. Examples include manufacturing workers, job setters, production control personnel, and typists, just to name a few.
- Methods: the methods include "how things are done," or in other words: the methodology. Some examples can include directions for cleaning up a laboratory, methods for converting crude oil into gasoline, written procedure for making textbooks, and operating instructions for setting up computers in an office department.
- Machinery: inclusive in this category are the machines and equipment that are used in the process. Some examples include numerical control machines, computers, and punch presses.
- Materials: this represents the materials needed for transformation to a product or service. Examples include steel, rivets, screws, typewriter paper, treated lumber for landscaping, and concrete mix.
- Monies: this category consists of the financial resources needed to operate the process.
- Message Media: this category represents the advertising and other communication mechanisms necessary to market the products or services.

Not every process has to possess all of the Ms. For example, there is no machine category in the production of handmade lemonade at a lemonade stand at the street corner, but there is the machinery category in the process of baking cakes (for example, ovens). Processes are all unique and with this uniqueness comes the fact that not every process uses the same Ms or the same elements within that particular M category (for example, not every process uses steel in the materials category; not every process uses punch presses in the machinery category; and not every process uses job setters).

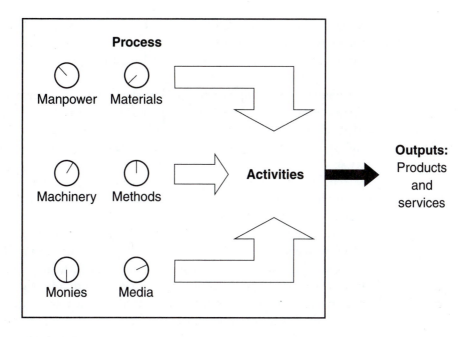

**FIGURE 4–1**
Process Concept

The conceptual diagram of a process is shown in Figure 4–1. The *M*s are shown as the dials of the process and are integrated with the activities of the process to yield an output in the form of products and services.

This concept of *M*s as dials reappears throughout the text in various portions of the quality recipe (for example, Design of Experiments). This is because the *M*s can be adjusted in terms of quantity and quality. For example, the quantity of manpower associated with a particular process can go from one to a large number of people. The quality of manpower can go from people hired right off the street to highly skilled and trained personnel within the company who are intimately familiar with the machinery and processes. The machinery can be very basic to extremely technical. And the materials can go from the lowest in price to very high in price.

---

**EXAMPLE 1**

    – Consider a pharmacy located in a very large hospital.
    – The pharmacy services patients, employees, and the local area residents.
    – The pharmacy process delivers an output: the product being filled is prescriptions.
    – Some of the *M*s of the process are as follow:

Manpower: Four pharmacists; two clerks for stock chasing, making labels, and taking payments; three administration people for in-processing, computer checks to ensure that the prescription is legitimately authorized by the doctor, and record keeping.

Machinery: Three computers hooked into the hospital mainframe network; two laser jet printers, and one pill dispenser possessing over 100 slots for pills.

Methods: Pharmacy department operations handbook inclusive of procedures and policy. The pharmacy operates seven days a week between 7 A.M. and 9 P.M.

Materials: Prescription medication in pill, ointment, and liquid form; printer paper for creating billing labels; plastic bottles for pills; and small jars for ointments and liquids.

Monies: The pharmacy possesses a $750K per year budget.

Message media: Pamphlets are regularly placed in various departments throughout the hospital, as well as in the individual patient rooms, informing patients of the in-hospital pharmacy services. In addition, when patients are out-processed from the hospital, the attending nurse or volunteer verbally informs the patient of the pharmacy services.

– The *M*s of the pharmacy department are interconnected via a set of interrelated activities to produce an output.

---

In continuing with the concept of processes, the picture of the process is extended as shown in Figure 4–2.

In the diagram, it can be seen that suppliers and vendors supply the materials for the process. The customers receive the finished products and services produced by the process. There also exists a very important element: feedback loops. These feedback loops provide valuable input as to whether the specific process is delivering quality products and services as well as whether incoming material from vendors is of quality grade.

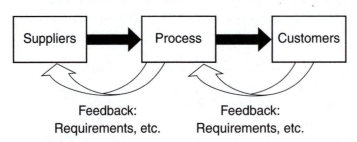

**FIGURE 4–2**
Extending the Process Concept

# 4.5 The Process Checklist

It is extremely vital to understand processes before moving forward with process improvement initiatives. Many times white and blue collar processes are arbitrarily **"tweaked"** in the name of process improvement. And many times, these "tweaks" are nothing more than the implementation of a perceived good idea. Because these tweaks are perceived as a good idea, many times the idea may not have been carefully thought out or planned as proposed by the Deming Plan/Do/Show/Act Cycle. Rather, they are sometimes cleverly disguised and sold to the decision makers as the ultimate process improvement initiative. Although there is nothing wrong with process improvement, there is something drastically wrong with making arbitrary changes that do not stem from some form of analysis, such as the **process checklist**, which is discussed in the next section.

## 4.5.1 Part I of the Process Checklist

To prove the point of the importance of understanding the process before making any change, review the questions in Part 1 of the Process Checklist, which is shown in Table 4–1. Part 1 of the checklist is used to understand the process via documentation before making any changes. In looking at the questions, it is very logical and correct to assume that anyone who is making a change to a process should be readily able to completely answer and document all answers before process improvement.

**TABLE 4–1**
Part 1 of the Process Checklist

---

- What is your job?
- Into what process does it fit?
- What are the specific tasks of this process?
- Who owns the process?
- Has the process been modeled or flowcharted?
- What departments does the process cross or affect?
- What products or services result from the process?
- Who are your internal and external customers?
- What are the specific and measurable customer requirements?
- How effective is the process in giving the customer the right thing right the first time?
- Who are your internal and external suppliers?
- How efficient is the process in terms of the expenditure of resources, rework, breakdowns, and delays?
- Has the process been analyzed? If so, how?
- What is the process measure of central tendency and variation?
- Has the process been baselined?

---

**TABLE 4–2**
Part 2 of the Process Checklist

- How did the improvement quantifiably improve the process?
- What was the impact on the process centering and variation?
- What was the impact on vendors and suppliers, if any?
- What was the impact on efficiency?
- What was the impact on internal customers (if any)?
- How will the improvement be monitored and documented?

If the answers to the questions in this checklist cannot be answered, then it is not certain that the process has even been totally understood. A complete understanding of a process is the prerequisite before changes to the process are made. After the process has been completely understood, then a process improvement usually takes place. When that happens, Part 2 of the Process Checklist must be answered and documented in order to ensure that the process has not been adversely affected. Table 4–2 depicts Part 2 of the checklist.

The process checklist shown in Table 4–1 and Table 4–2 is addressed again in the roadmap of process improvement, but we address it here to show the vital importance of understanding processes when addressing any changes or improvements.

**EXAMPLE 1**    In Chapter 1 (Section 1.8.2, Example 1), Angelica Fraer was introduced. If you recall, Angelica was challenged to improve her bookstores' productivity numbers and show results in a relatively short period of time (three months). Angelica reacted by implementing five new process changes that she strongly felt would improve her productivity numbers. Angelica implemented these changes in a vacuum without respect to the Plan/Do/Show/Act cycle philosophy or without even considering the process checklist shown in Table 4–1 and Table 4–2. Recall that her implemented changes included:

1. Laying off one clerk and spreading out the laid off clerk's workload to other employees as additional duties
2. Beginning a permanent-part-time initiative for her workers (a buzz word Angelica heard at a management conference. The concept is to replace full time workers with permanent part-time workers in an effort to reduce the amount of benefits a company has to pay)
3. Cutting store supplies across the board by 15% per month

4. Lowering the thermostat setting in the store by 10° in an effort to reduce energy costs consumed by the store

5. Deciding to reduce her employees' revenue sharing commission amounts by 2.5% in addition to putting them on permanent-part time

In considering the facts of the case presented in Chapter 1, Angelica's initiatives can be laid next to the process checklist. It will then be interesting to note how many of the questions could actually be answered before or after any of the changes were implemented. The answer is relatively obvious: not many; at least not many of the tougher questions (for example, efficiency, centering, and variation).

It is also important to keep in mind the lasting negative impact that will occur from not completing a process checklist. Remember that Angelica was rewarded for her improved performance and promoted to handle new responsibilities. Her successor was left with not only the task of understanding the stores' processes and his responsibilities but with understanding any longer-term ramifications resulting from Angelica's initiatives. The audit trail via the process checklist was vague, if not completely missing.

**EXAMPLE 2**

– In a hospital the management (with the help of quality consultant group) decided to implement a process improvement initiative that was intended not only to improve the hospital processes but to supposedly improve and emphasize the hospital's commitment to service.

– One of the initiatives was to put an "Idea Box" in each department such as the pharmacy, emergency room, admitting, and pediatrics. The employees were encouraged to submit their ideas. These boxes were further advertised as a method for implementing employee empowerment as relating to the improvement of the hospital's services. Monthly contests between departments were held, and the department with the most implemented suggestions received a decorated cake for the employees to share. Buttons, pins, and certificates were awarded to the employees when their recommendations were implemented. "Metrics" showing the number of new ideas implemented by departments and for the overall hospital were proudly displayed at the main entrance to the hospital.

– The entire improvement program was promoted through the staff via flyers and motivational memos passed throughout the rank. In addition, the CEO of the hospital occasionally made surprise visits to various departments to back-slap and encourage workers to "Keep up the great ideas!"

- A huge number of ideas were submitted for this novel approach to process improvement. Some ideas were good but were made without consideration to impacts on the processes, let alone answering any of the questions in either part of the process checklist.
- The feeding frenzy associated with the Idea Box soon began to shift focus to a concept primarily concerned with "How can we save money?"
- A sampling of some of the ideas and justifications include:

  • Personnel and supply cuts resulted because ". . . if this idea is implemented, we can save 1 full-time equivalent (FTE) in this department."
  • Janitorial service reductions occurred because "We can clean the main reception areas with a company that doesn't charge as much" and "We can vacuum the surgery waiting and preadmission testing areas once every other day as opposed to every day. This will save money!"
  • Elimination of employee morale picnics would definitely save money.
  • Permanent part-time employees over full-time employees (especially in the nursing staff ranks) became a way of life in an effort to ". . . reduce the costs associated with paying full-time benefits."
  • "Senior volunteers can wheel the patients out to their cars instead of nurses. This will free up the nurses to perform other more important duties."

- Money was being saved (at least in the short term) with absolutely no regard to long-term survivability, the processes, quality, or the customer.
- After the dust had settled, after the newness of all the buttons and pins wore off, after all of the backslapping and applause, after the consulting group cashed their check, and after healthy bonuses for members of management were handed out, the hospital declared success! They proudly boasted that they had now arrived at effective process improvement, and that the quality of the hospital's services had now reached an all-time high!
- Reality set in. The sacrifice of quality by not even analyzing processes via the process checklist before making changes was soon felt. Employee morale became terrible because some of the changes that called for reduced manpower began to overload some sections of administration and some departments in the nursing arena, especially in peak times. The overall cleanliness of the hospital deteriorated because the "new and less expensive" janitorial service was good at first but continued to gradually provide reduced vacuuming and cleaning services in the high-traffic areas such as surgery-waiting and in-processing areas. Supplies were not always available. And, unfortunately, some of the schedules were impossible to meet, with some of the new personnel cuts (or personnel shifts as management called it), that some patients did not receive their lunches until 2:30 in the afternoon. Chaos began to abound. No longer could the top management be seen walking around

the floors of the hospital backslapping employees and encouraging them to submit ideas and thanking them for "Working so hard!" Extremely negative comments began to be heard about management and the hospital both from employees in the hospital as well as from patients of the hospital.

---

This case demonstrates not only the criticality of ignoring the value of the process checklist, but demonstrates a complete lack of understanding of any of the Dr. Deming, Dr. Juran, or Philip Crosby points on quality. There was a complete failure of grasping even one aspect of their philosophies or any of the tenants of quality. After the first year, the cold hard truth slapped the hospital in the face:

- For the first time in the hospital's history a five-month decline in the number of patients was experienced in the pediatric, physical therapy, and maternity departments.
- The number of grievances was on the rise over the last seven months.
- Not only was the number of applications received by the personnel department down (although the hospital management refused to believe that it had anything to do with the permanent-part-time program), but the average years of experience for new applicants for hire was 1.6 (the hospital was well known and marketed for the experience it brought to the medical community).
- And monthly profits were down for nine consecutive months.

## 4.5.2 Other Process Aspects for the Checklist

As emphasized in the previous sections, the usage of a process checklist is very important. In addition, there is nothing wrong by including some of the following considerations in Part 2 of the process checklist to highlight the impacts of process changes and make the checklist more exhaustive:

- Process Layouts or how the process is physically laid out is a desirable item to include in the process checklist. Layouts of the product-oriented type (where facilities lie along product lines) or the process-oriented type (where the facilities lie in groups or batches of unique jobs or specialities) are just a couple of layout types.

  CHECKLIST QUESTION: Will the process change make an impact on the current layout or require changing the current layout?

- Process Resource types of considerations are also a desirable process checklist feature. Material movement distances, special material handling equipment, and the cost associated with new layouts are all factors in this category.

CHECKLIST QUESTION: Will the process change make an impact on the resources or require additional resources?

- Process Technology Upgrades definitely need to be included in this touch-of-a-button society. Upgraded computers, special software programs, and new automated equipment are just a few of the items in this category.

CHECKLIST QUESTION: Will the process change require any replacement of equipment or upgrade to computer software to accommodate it?

- Management Support should also be considered in the checklist. It must be remembered that management is charged with setting and determining the *M*s of a process, from how many personnel to how much money will be budgeted. As Dr. Deming states, management is ultimately responsible for the process. For example, in our pharmacy in Example 1 of Section 4.4, no worker in the pharmacy can come into work one day and automatically change the operating hours of the pharmacy; no worker can automatically change the procedure to eliminate billing notices; no worker can automatically change the types of medication the pharmacy carries. Although the personnel of the pharmacy department might have process improvements to suggest, it is still the ultimate responsibility of the management to set the *M*s.

CHECKLIST QUESTION: Do you have management support for the process change(s) recommended, especially in terms of cost?

## 4.6 Other Process Considerations

By now the concept of processes and the importance of understanding the process through means of the process checklist should be apparent. Processes are not to be considered lightly because they are the means for producing throughput. There are also a couple of considerations associated with processes that need to be addressed at this point in time: process tasks and process measurements.

### 4.6.1 Process Tasks

The discussion thus far has centered around the resources associated with a process. The other half of the process definition has to do with the tasks of the process. An excellent way to look at the tasks is via a flowchart. Although there are many types of flowcharts, which will be discussed in detail in Chapter 5, a **flowchart** is defined as a graphic representation of all of the major steps (or tasks) of a process.

---

EXAMPLE 1    In returning to the pharmacy department example (Section 4.4, Example 1), a simple flowchart of the process tasks is shown in Figure 4–3.

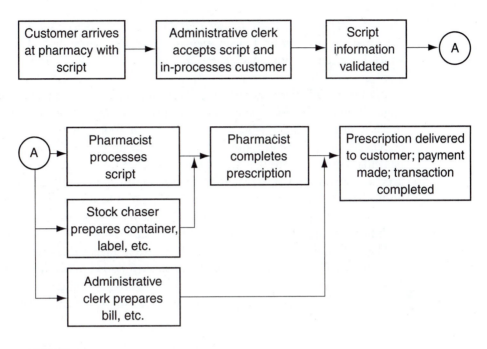

**FIGURE 4–3**
Pharmacy Flowchart

The flowchart demonstrates the logical ordering of the tasks associated with a process. It gives definition to the process in terms of boundaries and task relationships so that optimal decisions can be made.

---

**EXAMPLE 2**    In the pharmacy example, imagine that it was determined that the process needed to be improved by speeding it up. Consider how the process could be improved with respect to time if a flowchart did not exist to lay out the tasks that make up the process and their relationship to different tasks. Understanding all of the process tasks is important as well.

---

### 4.6.2 Process Measurement

Measurement, which was initially addressed in Chapter 2, is now addressed as it applies to processes. As has been seen, processes need to be measured to see if they are performing as expected, if they are efficient, or if process improvement can or needs to take place. To automatically assume that processes are efficient and never need to be measured is a gross error. Process reengineering is an entire field of study by itself, which is concerned with reducing the amount of waste, or inefficiency, associated with a process.

In addition, it is important to remember that process measurement is not applied in a quick and dirty manner. Knowing what specifically to measure in the process, how much to measure, and at what time increments to measure are important parameters associated with process measurement that must be understood. There are many companies and organizations that are deeply entrenched in the "metrics" syndrome where process measurement is the latest buzzword and measurements are gathered by the bushel basket with no regard as to what they mean. Quantity of metrics is sometimes perceived as being more important than the quality of those metrics. Metrics that do nothing more than count may not be of benefit in driving appropriate behavior.

The application of the process checklist and the Plan/Do/Show/Act cycle greatly assist in the measurement of processes. The quality roadmap, which begins in Chapter 5, takes the concept of process measurement and links it with customer requirements to produce sound information for effective process improvement initiatives.

## 4.7 Summary

- A process is a set of usually sequential value-added tasks that use organizational resources to produce a product or service.
- A process consists of what is known as the *M*s of a process: manpower, methods, materials, monies, media, machinery, and monies. Likewise, because processes are unique, not every process will have the same *M*s or the same elements within those specific *M*s.
- Vendors, or suppliers, provide inputs to the process in the form of material, etc. Customers receive the outputs from processes in the form of goods and services.
- It is vital to understand the many details associated with processes before injecting process improvement ideas that may or may not have been tested and proven. The Plan/Do/Show/Act cycle and process checklist are excellent techniques to ensure process understanding and a systematic approach to process improvement.
- A process checklist can be enhanced to include as many aspects of the process as desired. The more inclusive the checklist, the less turbulence will result when the process is improved. Other process checklist items include:

  ➢ Process Layout: how the process is physically laid out.
  ➢ Process Resources: considerations such as material movement distances, layout costs, and special material handling equipment.

➤ Process Technology: upgraded computers, special software programs, and new automated equipment.

➤ Management support is a vital aspect that should likewise be included in the process checklist. Because management is ultimately responsible for the process, including the financial resources, it is imperative that they be in the loop. This aspect of management should definitely be included in the checklist.

➤ A flowchart is a graphic representation of all major steps, or tasks, of a process. Note: Flowcharts are discussed in great detail in Chapter 5.

## Review Questions

1. Upgraded computers and specially developed software programs are considered:
   a. materials
   b. methods
   c. a process technology consideration
   d. manpower
   e. none of the above

2. The tasks of a process:
   a. must always be sequential
   b. may or may not be sequential
   c. must never be sequential
   d. none of the above

3. The best answer as to what feedback loops provide is:
   a. an additional group of taskings that can be implemented if the process delivers bad quality
   b. an alternative to process measurement
   c. inputs as to the methods associated with the process
   d. inputs as to the process quality and supplier quality
   e. c and d
   f. none of the above

4. _____ is/are ultimately responsible for the process.
   a. The workers
   b. Management
   c. Suppliers
   d. Customers
   e. All of the above

5. Process-oriented and product-oriented refer to:
   a. process layout
   b. methods
   c. materials
   d. process resources
   e. none of the above

6. Process tasks are best displayed:
   a. on a flowchart
   b. in a process checklist
   c. in the Plan section of the Plan/Do/Show/Act cycle
   d. none of the above

7. A list of detailed questions regarding specifics of a process is referred to as a:
   a. flowchart
   b. Plan/Do/Show/Act cycle
   c. process checklist
   d. none of the above

8. An example of a method is:
   a. the foreman of a department
   b. the operations manual
   c. the punch press
   d. all of the above

9. Typical questions that will probably appear in a checklist include:
   a. What is your job title?
   b. What is your job?
   c. Into what process(es) does your job fit?
   d. all of the above
   e. a and b
   f. b and c

10. Rivets, bolts, and screws are best classified as:
    a. materials
    b. machinery
    c. methods
    d. manpower
    e. none of the above

11. Production line workers and job setters are best classified as:
    a. materials
    b. machinery

    c. methods

    d. manpower

    e. none of the above

12. Contents of process flowcharts and training manuals are best classified as:
    a. materials
    b. machinery
    c. methods
    d. manpower
    e. none of the above

13. Punch presses and fork lifts are best classified as:
    a. materials
    b. machinery
    c. methods
    d. manpower
    e. none of the above

14. A graphic representation that demonstrates the logical ordering of tasks is a:
    a. process checklist
    b. flowchart
    c. Plan/Do/Show/Act cycle
    d. none of the above

15. Use of a process checklist before making changes to a process:
    a. will serve as an audit trail for process improvement
    b. documents the changes being made to the process
    c. identifies the Ms of a process
    d. none of the above

16. A set of tasks that use organizational resources is a:
    a. flowchart
    b. process checklist
    c. process
    d. none of the above

17. A sample of a feedback loop is:
    a. from the customer to the customer service representative
    b. from a foreman to a purchasing agent
    c. from a purchasing agent to a vendor
    d. all of the above
    e. none of the above

18. "Secretary duplicates letter for everyone on a distribution list" is an example of:
    a. an input to a process
    b. a process task
    c. a feedback loop
    d. none of the above

19. "The number of defective parts per shift" is an example of a:
    a. process task
    b. process output
    c. process measurement
    d. none of the above

20. Customer survey responses regarding product performance is an example of a:
    a. feedback loop
    b. process output
    c. process task
    d. none of the above

# Problems

1. Consider the process associated with a dry cleaning store where you are the store manager. The process starts with a customer dropping off some shirts to have them cleaned and pressed, and ends with the customer paying the bill and exiting the store. Although you may not be familiar with all of the micro tasks of a dry cleaning store:
    a. write the tasks you think would be needed in sequential order
    b. list the $M$s of the process and what elements are associated with each $M$ (for example, Materials: laundry soap)

2. For the dry cleaning store in problem 1, complete Part I of the process checklist. Note: for questions you may not be able to answer (for example, efficiency) describe what you would do to determine the answers.

3. The owner of the store has come in and asked you to determine an improvement that would speed up the operations. With this in mind:
    a. Describe your recommended improvement.
    b. Considering your improvement, answer the questions in Part II of the checklist. Again, if any question cannot be answered, describe what you would do to answer the question. Likewise, include in your answer the recommended questions for Part II such as layout and management support.
    c. What process measurements would you put in place?

4. Think of a process that you are familiar with. The process can be a personal process or one that you are familiar with at work. For that process:
   a. Describe the process.
   b. Write the tasks in sequential order.
   c. List the *M*s of the process and the elements associated with each *M*.
   d. Complete Part I of the checklist, answering the questions you can and describing what you would do to answer questions you do not know the answer to.

5. For your process in problem 4:
   a. Describe an improvement to the process.
   b. With your improvement in mind, answer the questions in Part II of the process checklist, inclusive of layout, management support, etc.
   c. What process measurements would you put in place?

# PART
# II

## *The Roadmap*

### Overview: Chapter 5–12

In the next section of this text, a roadmap for obtaining quality via process improvement is documented. The steps to the roadmap are sequential. The roadmap is not only designed and built on the concepts discussed in Chapters 1 through 4, but the roadmap is designed to build on itself, as each step builds on process data and information obtained from the previous step.

The roadmap specifically consists of the scientific techniques that comprise the world of quality and process improvement starting with the identification of a valid and specific customer requirement. The customer requirement step sets the stage for the remaining steps: flowcharting, Pareto and Fishbone analysis, statistical process control, process capability and process capability indices, and design of experiments. Following the roadmap will result in documented process improvement and improved product quality.

# 5

# Step 1: Customers and Customer Requirements

## 5.1 Chapter Objectives

After completing this chapter, the student should be able to:

- Understand the theory behind customer requirements as well as some of the key concepts associated with customer requirements.
- Evaluate a customer requirement to ensure it is specific and measurable.
- Gather process data to evaluate whether the process is delivering a quality product or service with respect to the customer requirement.
- Determine the amount of scrap or rework, if any, when comparing the process to the customer requirement.
- Document the appropriate information to baseline the process.

## 5.2 Key Terms

External Customer          Internal Customer          Requirements

## 5.3 Introduction

The purpose of this chapter is to begin the roadmap for quality by starting with the first step: the customer requirement. After the customer requirement is defined in specific and measurable terms, measures of central tendency and variability are applied to the process to:

- Determine factually whether a quality product or service is being delivered to the customer.
- Give insight into whether the process is a candidate for process improvement via variability reduction, central tendency shifting, or both.
- Gain a factual perspective into the percent of scrap, rework, and reinspection generated by the process.

It is extremely important to understand that there are two types of customers: **internal customers** and **external customers**. An internal customer is the next person in the organization who receives your product or service, whereas an external customer is the recipient of the final product or service. It is important to realize that even though most of the time companies are concerned with external customers, the internal customer is important as well. Internal customers are also in the production process for the delivery of quality products in a timely manner. The techniques described in the roadmap can be applied to both internal and external customers.

The beginning recipe steps covered in this chapter are shown in Figure 5–1.

---

*Roadmap step 1*

Step 1A:  Obtaining a valid and reliable
          customer requirement

Step 1B:  Gathering process data

Step 1C:  Determining the amount of
          scrap and rework from the
          process

Step 1D:  Processing documentation and
          baselining

---

**FIGURE 5–1**
Roadmap Step 1

## 5.4 Concepts Associated with Customers

The first and most vital concept associated with customers is that they and they alone are responsible for a company's throughput. Customers alone determine, through their purchase of the company's products and services, whether the company will gain or lose market share. Customers alone are the ultimate determinant for the success or failure of a company. And the customers alone constitute the Dr. Juran term of "society" that demands a company's products and services. It is very hard to imagine that any company can thrive or continue to thrive in a local or global economy when its products and services are not consumed. Customers possess power that companies must succumb to by producing the quality of products that customers demand.

Customers have choices when companies do not produce the quality products and services that they demand:

- They can leave and wait until the company produces the product they want.
- They can buy from another producer temporarily until the company produces what they desire.
- More often than not, when dissatisfied customers leave they seldom come back.

The customer message sent to companies that refuse to deliver quality goods and services is a powerful one: "I can always find someone who will take my money and give me what I want." The ultimate power of a customer is supported by the estimate that a happy customer tells five people while an unhappy customer tells at least fifteen.

The penalty for not giving customers their specific requirements usually results in rework, lost business through word of mouth, and complaints to local consumer agencies and watchdog groups, just to name a few.

A second concept is that most of the time customers know what they want. They do not need a spokesman to interpret their requirements. Customers are perfectly capable of stating exactly what they want. Not all customers may be able to convert their requirements to exact engineering specifications but they do know what they want. A perfect and quite simple example that the author has used endlessly in academic, industry, and government lectures is demonstrated in the following case.

### Student Exercise 1

- The audience is told that they are going to form a company—more specifically, a company that produces paper airplanes that the lecturer will buy for $1 per airplane.
- Half of the audience will be designers and manufacturers of these paper airplanes.
- The other half of the class will represent the ILITIES group (reliabILITY, qualITY, maintainabILITY, and sustainabILITY personnel). The ILITIES personnel can give

input to the designers and manufacturers as long as they do not get in their way (after all, the company is in business to manufacture and sell paper airplanes). Rather, it is more important to check and audit the product after manufacture to ensure that quality has been built in. NOTE: the lecturer makes various statements or comments like "We are in the business to manufacture" and "Will the ILITIES personnel please try to stay out of the way of production," which are counter to the Deming philosophy but may indeed exist in various companies.

— Each designer will have one sheet of paper and will each make one paper airplane.
— An industrial engineer is hired and given a stopwatch to make sure that we meet the standard of thirty seconds per airplane. After all, standards are important and must be adhered to in order to meet efficiency goals. And because we are in business, it is very important to meet various financial goals, one of which is to make paper airplanes efficiently (this is another statement made to point out a paradigm that sometimes exists in industry).
— After 30 seconds the designers will get up with their airplanes, stand behind a line, and fly them. The ILITIES personnel will evaluate the products.
— The lecturer then starts the exercise by saying "Start production!" The industrial engineer starts the stopwatch and the designers and manufacturers start to make paper airplanes.
— After 30 seconds the designers and manufacturers are stopped by the industrial engineer. The lecturer instructs the designers and manufacturers to take their products and stand behind a line and prepare to fly them one at a time. The lecturer takes out of his pocket twenty $1 bills, lays them on the podium and states that he is prepared to give each designer and manufacturer $1 for their airplane as soon as he sees that the product can meet his requirement.
— Before the airplanes are flown, the lecturer injects the fact that he wants the designers and manufacturers to demonstrate their airplanes and fly their airplanes so that they land on a small table over 20 ft. away.
— One by one, the designers fly their planes and one by one they all fail (most cannot even go 10 ft., let alone the required 20 ft.).
— The lecturer then takes a sheet of paper, crumples it up into a wad, throws it to the table and tells the audience that "This is what I wanted . . . this is my idea of a paper airplane. . . . I don't think I'll buy any of them."
— The points are made quickly and effectively:

  • Everyone automatically went out and made a paper airplane. Everyone knew what a paper airplane should look like. Everyone knew they could deliver that product. Everyone went out and, based on their own paradigms, built what they felt the customer meant by "paper airplane." Never has anyone ever come up to the lecturer and asked anything specifically about what is meant by a paper airplane, dimensions, or critical criteria.

- All products vary. No two airplanes look alike, no two airplanes perform the same, no two airplanes exhibit the same mean time between failures on critical parameters.
- The amount of rework, reinspection, analyzing, and checking required to bring any or all of the planes to a satisfactory status in order for them to be purchased would have a severe impact on the product line profit figure.
- Ultimately, what started out as a potentially very profitable enterprise, turned into a very costly adventure. The bottom line is the customer did not buy any planes. The customer will very probably not come back in the future, and the company's throughput is "zero."

A third concept that it is becoming more and more recognized and accepted as common knowledge is that meeting and exceeding customer requirements is the foundation on which successful companies reside. Recall from Chapter 1 that the Malcolm Baldrige Award is broken into seven categories of which "Customer Focus and Satisfaction" is one. Also recall that this category is worth 250 points (of these points, a little more than one third (90 points) deal with the approaches or processes that keep customers satisfied, while approximately two thirds (160 points) deal with customer satisfaction results. Only one other category in the Malcolm Baldrige Award is worth 250 points (Business Results). The rest of the categories are worth no more than 140 points, which further demonstrates the recognized significance of customers and their requirements.

When thinking about the concepts just discussed, it can be seen that it is imperative that the customer be listened to very closely. Too many times companies and their representatives (from automobiles to electronics, from cosmetics to real estate agents, and from bankers to service representatives) possess an air of superiority when dealing with customers. This attitude almost takes on an air of "how lucky you are to buy our products or services." This attitude also sometimes carries an erroneous belief that they actually know more than the customer. They fail to recognize from history that companies and their market share are affected by customers (the automotive and electronics industries are just two examples that have proven this point). Too many times companies do not realize or believe that consumers will and can go elsewhere; they are not bound by law to buy from any particular company. Too many times companies do not believe that a consumer "can find someone else to take their money." Too many times customer returns or complaints are just shrugged off without realizing that these complaints and returns not only cost the company money, but can be the basis of valuable inputs to the producers on how to improve their products. All too often, customer complaints and warranty problems are many times broken down and categorized into two viruses: "noise" problems and problems that result in economic expenditure by the company to correct. Too many times customers and what they have to say are just ignored. Unfortunately, too many times the impacts associated with these types of attitudes are devastating on the business. By the time management realizes that their customer base is deteriorating, it is usually too late because the customers are out the door, seldom to return.

# 5.5 Roadmap Step 1: Customer Requirements

The initial step of the quality roadmap is customer requirements. But what constitutes customer requirements? Is it producing a product and offering it to a customer with the hopes that it will be purchased? Is it producing a product and then asking various customers how this product can be improved to meet their needs? Customer **requirements** are critical and there is a specific method to ensure that the customer requirement obtained is a valid requirement. Likewise, there is a specific method to check that the process is delivering a quality product or service to the customer. These concerns are addressed in the very first step of the roadmap.

## 5.5.1 Step 1a: Obtaining a Valid and Reliable Customer Requirement

The first thing that needs to occur is that the customer needs to be contacted to determine his requirements. In many cases, such as the automotive and electronics industries, it is infeasible to contact "all" of the customers. That is where random sampling and surveying, which was discussed in Chapter 3, comes into play. Notwithstanding, the point to be made here is that the customers need to be contacted to determine exactly what they want. As stated earlier, neither the engineers, customer service representatives, CEOs, nor plant managers can determine the exact customer requirements. Only the *customer* can specify exactly what he wants. Recall the consequences demonstrated with Student Exercise 1 in Section 5.4 that occur when the posture of ". . . but we already know what the customer wants" is taken.

Although obtaining a valid and reliable customer requirement sounds simple, this requirement needs to satisfy two prerequisites: it must be specific and it must be measurable. The reasons for these prerequisites are:

- It is only through factual data and information, and quantification of those data and information, that the quality of products and services can be assessed.
- Without specificity and measurability, the techniques for quality and process improvement cannot be implemented. Statistical Process Control (SPC), Design of

Experiments (DOE), Pareto analysis, and Fishbone diagrams cannot be implemented with any degree of statistical confidence or success when customer requirements are not specific and measurable and translated back to critical parameters.

- The implementation of process improvement techniques requires data collection. Recall from Chapter 2 that data collection can be expensive and time consuming, and going back to collect "good data" to replace "dirty data" that were initially collected has invariably disheartened process improvement initiatives. Unfortunately, when this happens, managers and employees get a bad taste for quality initiatives, and these initiatives are then either abandoned or compromised in favor of suboptimal initiatives (for example, the "80% solution . . ." which in reality is merely an excuse for not doing the right thing right the first time).

Look at the following examples and evaluate each one of the customer requirements with respect to whether they are specific and measurable:

- A good tasting steak
- Light bulbs that have a mean time failure of between 2000 and 2050 hours
- A round tire
- A hula hoop weighing exactly 1/2 pound, with an inner diameter of 36.00 inches and an outer diameter of 36.75 inches
- A chemical mixture containing exactly 40.0% sulfur
- A pulley that can lift loading dock shipments
- Dial pads on telephones that measure 1/4 in. × 3/8 in.

Dr. Deming gave an example during his many seminars driving home the importance of well-defined customer requirements. He pointed to a table and said the customer requirement was to have a clean table. Dr. Deming said he did not know what a clean table meant. Did it mean clean enough to serve as a workbench? Or did it mean clean enough to write a report on? Or did it mean clean enough to eat dinner from? Or did it mean clean enough to perform surgery on? As seen with Dr. Deming's example, it is imperative to understand specifically and measurably what the customer requirement is.

Although not every customer can translate his requirements into exact design and engineering specifications, it is still incumbent upon the producer to ensure that the design and engineering specifications correctly reflect the customer's requirements. Some common questions that serve as a basis for developing customer requirements are included in the Customer Requirement Checklist shown in Table 5–1:

**TABLE 5–1**
Customer Requirement Checklist Questions

1. What is the product or service provided to the customer?
2. What characteristics are important to the customer?
3. Is each characteristic defined specifically?
4. Is each characteristic defined measurably?
5. What are some of the complaints against the product or service? (This can help expose requirements that are important to the customer.)

Figure 5–2 depicts the procedure for obtaining a valid customer requirement.

## 5.5.2 Customer Requirements Cavaets to Remember for Step 1A

CAVEAT 1: To this point the importance of customer requirements has been emphasized and reemphasized. Notwithstanding, a degree of caution must be exercised. There will always be a portion of the customer population who would like to purchase a new automobile for $500, purchase a complete sound system for $200, or in essence, receive some utility at below fair market value. When dealing with this type of

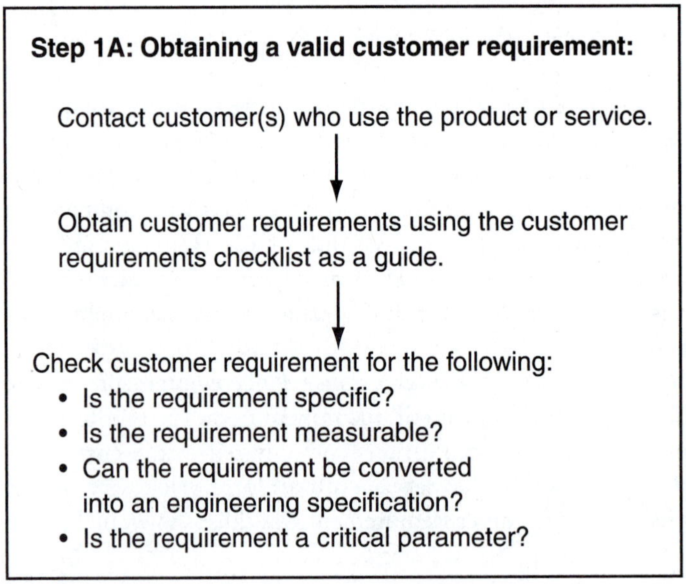

---

**Step 1A: Obtaining a valid customer requirement:**

Contact customer(s) who use the product or service.

Obtain customer requirements using the customer requirements checklist as a guide.

Check customer requirement for the following:
- Is the requirement specific?
- Is the requirement measurable?
- Can the requirement be converted into an engineering specification?
- Is the requirement a critical parameter?

---

**FIGURE 5–2**
Step 1A: Procedure for Obtaining a Valid Customer Requirement

scenario, the quality (giving the customer the right thing right the first time) becomes a moot point, especially because there will be no throughput. The opposite of this situation is that there will always be a producer who will feel compelled to charge prices in excess of fair market value. Likewise, in this case quality will become a moot point because there will be no throughput. As can be seen from these two scenarios there must be throughput in order to apply quality and process improvement techniques. Consider the following example as a demonstration.

---

**EXAMPLE 1**

- The lecturer asks the audience how many would be interested in buying an automobile that was guaranteed to never need any kind of maintenance for twenty years. Most of the audience raises their hands.
- The lecturer asks how many in the audience would pay $150,000 for this car. After a pause a hand or two go up.
- The lecturer asks how many in the audience would pay $100,000 for this car. A few more hands go up.
- The lecturer asks how many in the audience would pay $60,000 for this car. More hands go up, and although a majority of the hands in the audience are raised, it is still not unanimous.
- The lecturer asks how many would pay $30,000 for this car. Now most hands go up and just a few remaining audience members do not raise their hands. When singled out and asked why, several answers are given: "I like to buy a car every couple of years." "I want a car that I can play around with and work on and I would be afraid to mess around with a $30,000 car." "I will never pay more than $15,000 for any car."
- Finally, the lecturer lets the audience in on a hint: the cost to manufacture the automobile is $150,000 (not including overhead, profit margin, and other financial aspects).
- The points to be made in this case is that although the customer and his requirements have been emphasized there must exist a balance between the producer sales cost and the cost the customer is willing to pay. The producer may have the perfect product, but unless there is throughput, quality and process improvement are moot.

---

CAVEAT 2: Naturally, when there is more than one customer involved, the process of defining a valid customer requirement may be a little more complicated. No two people, or groups of people, want the same identical characteristics. Therefore, it may become necessary to segregate the population into target market groups (or groups that have a little narrower requirement than the entire population). Sampling the population with instruments such as surveys is one technique to determine different market groups and their different customer requirements. When thinking of the General Motors Corporation, the technique of different groups for different products is apparent. For example, General Motors has many targeted groups at various levels. At the top level is the targeted group that chooses to buy General Motors vehicles. Below that level are the different divisions (Chevrolet, Buick, Cadillac, etc.), each having different market groups. And within each division is a plethora of vehicles available for consumption by different markets. As an example of the different target groups, the 1980s Pontiac Firebird and Chevrolet Camaro were of the same body style ("F") and had similar but unique designs (for example, a Pontiac engine versus a Chevrolet engine).

In cases where there is more than one customer, it is very apparent that it is impossible to satisfy all customers. Companies usually focus on satisfying a majority of the customers who, in turn, buy the products and services of the company. Where there are many customers with varying requirements, it is beneficial to break down the customers into target groups. This allows for developing a more homogeneous requirement within a specific group.

### 5.5.3 Step 1B: Gathering Process Data

Once the customer requirement has been determined, it is now time to gather some initial data from the manufacturing processes to gain factual and quantifiable insight into:

- Whether the process is delivering a quality product or service
- Which aspect of process improvement (variability and central tendency shifting) needs to occur, if any

Recall from Chapter 1 that a good definition of quality was given by "Giving the customer the right thing right the first time." Recall also from Chapter 1 that quality was described as a science with very provable methods. It is at this point in time that the definition of quality is applied to process data, and the process data are subjected to mathematical or statistical concepts to see if the process is delivering a quality product or service. The definition of quality is rewritten again, only this time a little differently. The definition (Giving the customer the right thing right the first time) is divided into two parts for the linkage to the provable methods.

## Part One of the Quality Equation: Giving the Customer the Right Thing Right

This side of the definition has to do with the mathematical concept of central tendency. In other words, by using a measure of central tendency we will be able to answer if the customer is receiving the right thing right, which is part one of the quality equation. Stating in mathematical terms: giving the customer the right thing right is a function of central tendency. Furthermore, while there are several measures of central tendency (that is, median, mode), the average, or mean will be the mathematical expressions used to answer part one. In mathematical nomenclature:

*Quality Definition*                    *Mathematical Definition*

Giving the customer    ➡    central tendency

the right thing right

*where:*

- the measure of central tendency utilized will be the average (or mean)

---

**EXAMPLE 1**   Charlotte has a requirement: she wants to buy 1 in. diameter steel rods exactly 10.0 in. in length. She wants to buy 100 of these rods. The 10.0 in. length is extremely critical to her, and in this example it can be seen that her customer requirement is very specific and measurable. The type of steel and other parameters are not as of extreme critical interest to her. Note that the discrimination required in her requirement is to tenths of an inch (recall from Chapter 2 that 10 in. does not equal 10.0 in., which does not equal 10.00 in.). Charlotte goes to the local hardware store and places an order for the steel rods to be picked up the next day. The hardware store clerk goes back to the model shop, which is conveniently located in the back of the hardware store, and gives the machinist her order. The machinist sets up the machine and begins to produce the steel rods out of 1 in. diameter stock. Unknown to the machinist, certain tooling on the machine was beginning to wear at an abnormal rate (later chapters deal with the techniques to proactively predict when something in a process is beginning to deteriorate). In fact, last week several other customer orders were filled and there appeared to be no problem with those orders. Thus, the machinist did not hesitate to set up the job and begin processing Charlotte's order.

When Charlotte went in the next day she sampled ten from the lot and began to measure them to ensure that the rods satisfy her requirement. The measurements of the sample are shown in Table 5–2.

**TABLE 5–2**

Sample Data of Steel Rod Length

| | |
|---|---|
| 9.8 in. | 9.9 in. |
| 9.9 in. | 9.9 in. |
| 10.0 in. | 10.0 in. |
| 9.8 in. | 9.9 in. |
| 9.8 in. | 10.0 in. |

Table 5–2 shows that the average length of the sampled rods is 9.9 in. and this does not equal the 10.0 in. customer requirement. In terms of "Giving the customer the right thing right" it can be seen that the average gives us initial insight into this aspect, because the customer is not getting the right thing. Another way of stating it, for clarity, is that on the average it is apparent that the customer is not receiving the right thing right (9.9 in. does not equal 10.0 in.).

Taking the average is not enough, however. The remaining portion of the quality definition (the first time) has not been addressed and needs to be addressed to see the totality of the quality situation.

## Part Two of the Quality Equation: The First Time

In Example 1, Charlotte sampled ten products. Notice that she did not want one out of ten to be correct in length. She also did not want five out of ten to be correct in length. And she did not want nine out of ten to be correct in length. Charlotte wanted all ten to be correct in length. She wanted no variation. She wanted all of the steel rods to be the exact length she requested. In other words, she wanted all rods to be right the first time. Thus, part two of the quality equation has to do with variability. Remembering from Chapter 2 that there are different measures of variability (for example, range and standard deviation), one of those measures will answer part two of the quality equation. The range, which is the highest value in a data set minus the lowest value in the data set (Hi – Lo), tells whether the process is consistent:

- When the range is zero, then the process is consistent, not "consistently good" or "consistently bad"—just consistent.
- When the range is a number other than zero, then the process is inconsistent, not "inconsistently good" or "inconsistently bad"—just inconsistent.

In expressing part two as a mathematical expression:

| *Quality Definition* | | *Mathematical Definition* |
|---|---|---|
| the first time | ➡ | variability |

*where:*

– the measure of variability utilized will be the range.

---

**EXAMPLE 2**   Going back to Example 1, the highest value of Charlotte's sample was 10.0 in. and the lowest value was 9.8 in. Thus the range is 10.0 in. – 9.8 in. = 0.2 in. The range does not equal zero so the process is inconsistent, not inconsistently good or bad—just inconsistent. This inconsistency means there is variability in the process.

---

It is time to put the two parts of the quality equation (central tendency and variability) together to form the entire quality equation. Doing this determines if the current process is giving the customer the right thing right the first time. It is very important to point out that both of these components together determine quality, not just one or the other. Both components have to be used together to answer the question of whether quality is met. And as will be seen in later chapters, these components of average and range are used in control charts (the X-bar and R chart) for process monitoring, control, and improvement. This application of simple statistics to the process delivering the customer requirement is a step that gives valuable information to the manufacturer.

---

**EXAMPLE 3**   In continuing with the example with Charlotte, the total quality summary statistics are as follows:

| | |
|---|---|
| Customer Requirement: | 1 in. diameter steel rods with a length of 10.0 in. |
| Quantity: | 100 |
| Process Data: | A sample of 10% of the order yielded an average of 9.9 in. with a range of 0.2 in. |
| Interpretation: | The process is not giving the customer what she wants consistently (sometimes the rods are 10.0 in. and sometimes they are not). Process improvement needs to occur if the hardware store wishes to keep Charlotte as its customer. |

**EXAMPLE 4**   In a new plant, time cards need to be submitted to the payroll office within 30 minutes after the close of the last shift on Friday. After twenty weeks of operation, the payroll office complains to the plant manager that this process is not being adhered to and as a result the payroll checks are not being processed on time and workers are starting to complain. The foremen contend that they are delivering the time cards to payroll on time most of the time and that it is not their fault. Finger-pointing prevails. The payroll office then goes back and gathers some sample data (recorded in minutes), which are shown in Table 5–3.

**TABLE 5–3**
Sample Data of Payroll Submission Times

| | | | |
|---|---|---|---|
| 30 | 30 | 27 | 33 |
| 30 | 30 | 32 | 28 |
| 28 | 30 | 32 | 29 |
| 29 | 30 | 31 | 29 |
| 31 | 30 | 31 | 30 |

Now that the process data have been collected, the quality summary statistics can be demonstrated as follows:

| | |
|---|---|
| Customer Requirement: | 30 minutes |
| Process Statistics: | Average = 30 minutes |
| Range: | 5 minutes |
| Histogram of Performance: | The histogram is shown in Figure 5–3 |
| Interpretation: | The process is giving the customer the right thing right on the average (30 minutes), but it is inconsistent (range = 5 minutes). This inconsistency (process variability) is what is causing problems in the payroll department. In looking at the raw data, this process is not meeting the 30-minute requirement six out of twenty times, or 30%. Thinking about this, it can be seen that 30% of the time, rework, extra expense, and complaining customers are going to result. Now, armed with some simple process statistics, a variability reduction program needs to be implemented. |

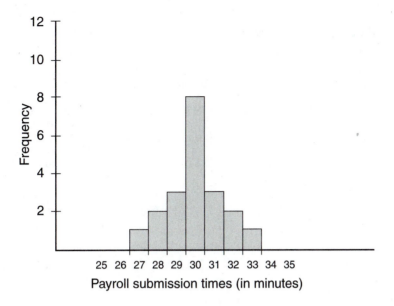

**FIGURE 5–3**
Histogram of Payroll Submission Times

---

**EXAMPLE 5**  A new engine chassis has been designed for use in a group of race cars that run 100-mile races every weekend. The car maintenance crew chiefs told the manufacturer that the chassis requires readjustment after each race (125 miles is the race distance, inclusive of testing and time trials) and the process is time consuming and costly. The car sponsors stated that if the new chassis lasted for two races (250 miles) before scheduled adjustments, they would buy them because they were much lighter than the standard chassis. However, the sponsors wanted to see some test data before committing to the new part and signing contracts. The manufacturer of the chassis put the chassis on twenty-five test race cars and gathered data on the total distance the chassis lasted before it required adjustment. The data (recorded in miles before readjustment) from the race cars were collected and are displayed in Table 5–4.

**TABLE 5–4**
Sample Data of Chassis Mileage Before Readjustment

| | | | | |
|---|---|---|---|---|
| 500 | 550 | 500 | 400 | 450 |
| 400 | 500 | 600 | 450 | 450 |
| 500 | 500 | 550 | 600 | 350 |
| 550 | 650 | 550 | 500 | 600 |
| 500 | 500 | 450 | 400 | 500 |

In the same manner as before, the quality summary statistics can be computed and interpreted as follow:

Customer Requirement:        250 miles

Process Statistics:          Average = 500 miles

                             Range = 300 miles

Histogram of Performance:    The histogram is shown in Figure 5–4.

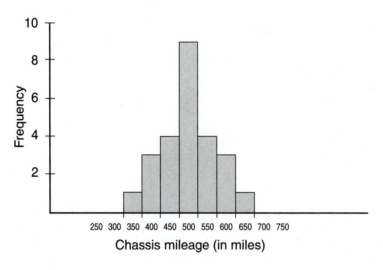

**FIGURE 5–4**
Histogram of Chassis Mileage

Interpretation:              As can be seen from the sampled data, the customer requirement is exceeded and although the process is inconsistent, the variability is not resulting in any chassis failing to meet the customer requirements. Notwithstanding, the manufacturer should still take note that variability exists and should incorporate some process improvement initiatives aimed at reducing variability.

It should be very evident from the preceding examples the power of quantifying the process with respect to the customer requirements. Imagine a meeting where the managers and workforce are trying to determine whether they need to begin a process improvement initiative. Throughout the room personnel are voicing their opinions: "The process gives us good parts most of the time." "No, I remember one time the process produced scrap." "Where do we start if we were to start an initiative?" "I vote for a product improvement team." "Why don't we contact some customers and see what they think of our product?" "Let's gather some data." Now amidst all of the discussion and opinions, one person stands up and states the following facts in a systematic manner: "The customer requirement is 10 kilograms ± 0.5 kilogram. A sampling indicates that the process mean is 10 kilograms but the process is inconsistent in that the range of parts produced is from 9.1 kilograms to 10.9 kilograms. This is indicative of a variability problem. We need to implement a variability reduction initiative." Imagine the impact this would have on the group. The previously expressed opinions would be greatly overshadowed by the facts obtained by applying Step 1B of the roadmap.

The quantification of a specific and measurable customer requirement is the first step in process improvement and is very important. It is the very basis for the other quantifiable techniques associated with quality that will be continuously applied to processes to obtain information that will result in documented process improvement and improved quality.

The procedure for Step 1B is shown in Figure 5–5.

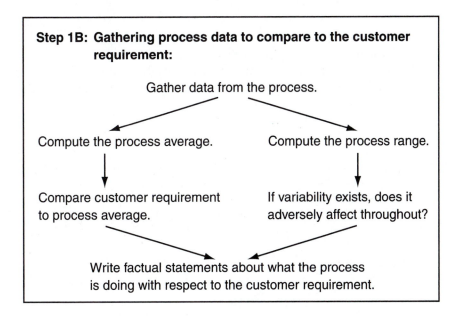

**FIGURE 5–5**
Step 1B: Procedure for Gathering Process Data

### 5.5.4 Step 1C: Determining the Amount of Scrap and Rework from the Process

Now that the answer of whether the process is giving the customer the right thing right the first time has been determined, further knowledge about the cost associated with not delivering quality can be obtained. As discussed in Chapter 1, the obvious costs that come to mind when quality is not achieved are scrap, rework, reinspection, possible overtime, and possible lost market share, just to name a few. Although many articles have been written trying to pinpoint this cost, it can be reiterated that the total cost is intangible. Notwithstanding, the roadmap can be continued to determine some of the apparent costs via application of the Z-test (reference Chapter 3) to the process data obtained in Step 1B. Recall from Chapter 3 that many processes are described by a normal curve, and that the standard deviation is a measure of the normal curve's variability. Also recall that the Z-test determines the area to the left of some data point. These concepts are in the next step of the quality recipe, and the best way to explain this step is through an example.

An extremely large volume customer requires that the lightbulbs have a minimum life of 985 hours. The criticality of this parameter is proven by the fact that the customer pays an "incentive" of an extra 25 cents per bulb to the producer for this expected extra high-quality level. The customer has registered a complaint that the lots he is receiving include too much scrap and that possibly he will be looking to another vendor to procure these bulbs. The manufacturer decides to obtain a sample of twenty-eight lightbulbs from a production lot and life test them. The life test results (in hours) are shown in Table 5–5.

In continuing with the procedure, the quality summary statistics are now computed, but note that the standard deviation is also computed and is shown under the process statistics portion as well:

**TABLE 5–5**
Sample Data of Lightbulbs

| | | | |
|---|---|---|---|
| 1000 | 1050 | 1000 | 950 |
| 1150 | 1100 | 1000 | 850 |
| 950 | 1000 | 1050 | 1050 |
| 1000 | 900 | 1000 | 950 |
| 1000 | 900 | 1000 | 1100 |
| 1000 | 1050 | 1000 | 900 |
| 950 | 1100 | 950 | 1050 |

| | |
|---|---|
| Customer Requirement: | Minimum life of 985 hours |
| Process Statistics: | Average = 1000 hours |
| | Range = 300 hours |
| | Standard Deviation = 69.4 hours |
| Histogram of Performance: | The histogram is shown in Figure 5–6. |

In looking at the histogram, a vertical line has been drawn to show the customer requirement. Using the Z value to determine the area to the left of 985 hours (or the percent of lightbulbs that are not of use to the customer):

$$Z = \frac{\chi - \mu}{\sigma}$$

where: $\chi$ = the point of interest
$\mu$ = the process average
$\sigma$ = one standard deviation of the process

$$Z = \frac{985 - 1000}{69.4}$$

$$Z = -0.216$$

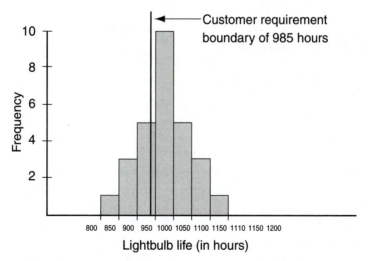

**FIGURE 5–6**
Histogram of Lightbulb Life

The corresponding value from the Z table for –0.216 is 0.4129 or 41.29%. This represents the area to the left of the minimum customer requirement of 985 hours. The process is yielding 41.29% unusable lightbulbs. In addition, the customer is rightfully upset, especially after paying an "incentive" for this type of quality. In looking at the producer aspect, he should quickly realize the following:

- Rectification of the shipped lots must occur.
- Until the process is improved the lots will need to be 100% inspected.
- A process improvement initiative needs to occur. Actually, had the producer implemented a quality-oriented technique such as statistical process control he would have been able to act proactively, as opposed to the now reactive situation he is in.
- The customer will in all likelihood revisit the "incentive" amount and renegotiate a new one, or use the producer's poor performance to eliminate the incentive.
- The intangible negative word of mouth publicity that the customer may speak to other potential customers could be devastating.

The usage of the Z tables to determine a rough estimate of scrap and rework is a valuable transition to the process improvement techniques detailed in the remainder of this text. It links the customer requirement to the process and further provides an initial estimate of how "good" the process is performing in relation to the customer requirement. The procedure for Step 1C of the roadmap is shown in Figure 5–7.

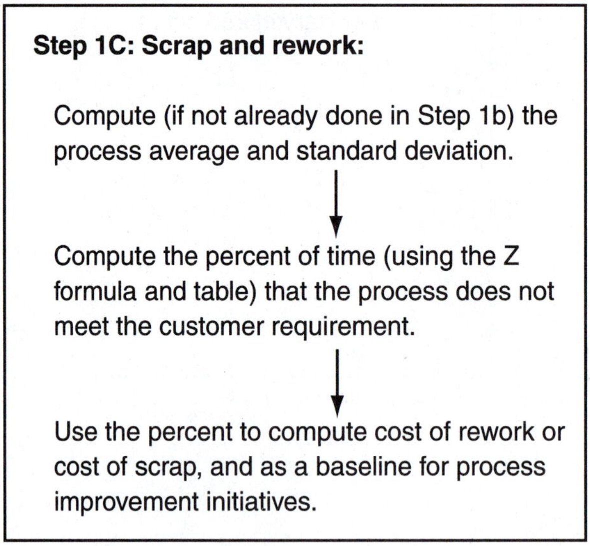

**FIGURE 5–7**
Step 1C: Procedure for Determining the Amount of Scrap and Rework

In addition, the information obtained from this step in the recipe serves as a baseline for process improvement initiatives. In this example, the initial scrap or rework percent from Step 1C is 41.29%.

### 5.5.5 Step 1D: Process Documentation and Baselining

The final item in Step 1 is that of documenting and baselining the process. From the first three steps (1A, 1B, and 1C), it can be seen that valuable quantifiable information has been gathered. It is this information that needs to be documented and serves as the baseline for the process. The documentation needs to encompass the quality summary statistics of the customer requirement, central tendency, variability, percent scrap or rework, and the histogram of the process.

As a simple example, suppose that the lightbulb producer in Example 1 of the previous section decides to implement a process improvement initiative, and that the percent scrap or rework is reduced to 28% after implementation of some initial proven improvements. The process improvement initiative can be quantified as reducing the scrap or rework by 13.29%. The 13.29%, as well as the measures of central tendency and variability, serve as the baseline for the new process. The documentation is not only necessary in answering ever present questions like "How much did we improve by implementing this new idea?" but serve as valuable information when applying for awards such as the Malcolm Baldrige Award. Likewise, the results of documented process improvement are invaluable in marketing and advertising a company's attitude toward delivering quality products and services to customers.

As will be seen in the remainder of the text, other statistical techniques are applied to processes to result in process improvement, and documentation, baselining, and rebaselining need to occur. After the initiative has been implemented, quantifiable statistics of the new process will be gathered, just as with this step. Without documentation, baselining, and rebaselining, it will not be known if various process improvements actually did improve the quality of the process or adversely affect the quality of the process.

## 5.6 Summary

- Delivering products and services that meet customer requirements is vital to the survival of any company (small, medium, or large), and it is vital that the exact customer requirements are known.
- Customers can be classified into two categories: internal customers and external customers.
- An internal customer is the next person in the organization who receives the product or service.

- An external customer is the recipient of the final product or service.
- Customers have choices if a company does not give them the quality product or service they are after. They can leave and wait until the company does produce what they are after; they can buy the product from another company, but return to the original company for future sales when the company produces what they want; or they can leave and never go back.
- Customers know what they want. Even though they may not be able to articulate the requirements into exact engineering specifications, industrial history has proven time and time again that they know what they want and they will usually only buy what they want.
- It is the customers (not the engineers, CEO, or anyone else for that matter) who specify what they want.
- The first step is to determine if the customer requirement is both specific and measurable. It is imperative that the customer requirement meet this test.
- The second step of the recipe is to gather data from the process to determine if quality is being delivered to the customer.
- Mathematical measures are applied to the data collected from the process to determine if the process is giving the customer quality. The mathematical measures and their relationship to the quality definition are:

  "Giving the customer the right thing right" is proven by the central tendency measure of average (or mean), whereas "the first time" portion of the quality definition is proven by variability measure range (Hi-Lo).

- Both measures of central tendency and variability must be used to determine if quality exists in a given product or service.
- The third step is to use the statistical concept of the Z-value to further extract process information with respect to scrap, rework, and reinspection.
- The first three steps provide the process statistics that serve as the baseline for the process.

## Review Questions

1. The definition of quality is:
   a. doing what the "boss" wants
   b. giving the customer the right thing right the first time
   c. the engineering department's specifications and designs
   d. none of the above

2. Which mathematical measure pertains to "giving the customer the right thing right"?
   a. Central tendency
   b. Variability
   c. Both a and b
   d. None of the above

3. Application of the Z-value to process data gives insight into:
   a. scrap
   b. rework
   c. cost of non-quality
   d. all of the above
   e. none of the above

4. An internal customer is:
   a. the end user of the company's product or service
   b. the next person who receives the product or service
   c. the next higher person in the chain of command
   d. none of the above

5. A customer requirement must be:
   a. specific
   b. repeatable
   c. measurable
   d. a and c
   e. none of the above

6. A tool that can assist with determining customer requirements is:
   a. the number of returned products
   b. customer requirements checklist
   c. engineering product improvement meetings
   d. none of the above

7. Which mathematical measure pertains to ". . . the first time"?
   a. Central tendency
   b. Variability
   c. Both a and b
   d. None of the above

8. The Z-test is used in which of the recipe steps?
   a. Step 1A: Obtaining a valid and reliable customer requirement.
   b. Step 1B: Gathering process data to determine how the process is performing with respect to the customer requirement.

    c. Step 1C: Determining the amount of scrap, rework, and reinspection.

    d. All of the above.

    e. None of the above.

9. Unhappy customers will usually tell _____ people about their unpleasant experience.

    a. two

    b. five

    c. ten

    d. none of the above

10. When quantifying quality, the most important mathematical measure is:

    a. central tendency

    b. variability

    c. both a and b

    d. none of the above

11. An external customer is:

    a. the end user of the company's product or service

    b. the next person who receives the product or service

    c. the next higher person in the chain of command

    d. none of the above

12. When variability exists in a process, it will:

    a. always adversely affect the product or service

    b. sometimes adversely affect the product or service

    c. never adversely affect the product or service

    d. none of the above

13. Happy customers will usually tell _____ people about their pleasant customer experience.

    a. five

    b. ten

    c. at least fifteen

    d. none of the above

14. A logical goal of Step 1A (Obtaining a valid and reliable customer requirement) is to:

    a. identify whether the process is giving the customer the right thing right the first time

    b. evaluate the percent rework that is being generated

    c. ensure that the manufacturing department can build the product called for in the engineering specifications

    d. none of the above

15. One of the goals of Step 1B (Gathering process data to determine how the process is performing with respect to the customer requirement) is to:
    a. calculate the standard deviation for use in the Z-test
    b. gain insight into which aspect of process improvement (central tendency or variability, or both) needs to occur to improve the quality of the product or service
    c. determine a specific and measurable customer requirement
    d. none of the above

16. Dr. Deming's "clean table" discussion pertains mostly to which step in the recipe?
    a. Step 1A: Obtaining a valid and reliable customer requirement.
    b. Step 1B: Gathering process data to determine how the process is performing with respect to the customer requirement.
    c. Step 1C: Determining the amount of scrap, rework, reinspection, etc.
    d. All of the above.
    e. None of the above.

17. The customer requirements checklist pertains only to:
    a. internal customers
    b. external customers
    c. both internal and external customers
    d. none of the above

18. The percent scrap and rework can be used:
    a. as a baseline for process improvement initiatives
    b. when the sampled items are not randomly selected
    c. even if the customer requirement is not specific and measurable
    d. none of the above

19. The customer requirement can be modified by the company:
    a. whenever the company does not have the manufacturing capability to produce the product or service that satisfies the customer
    b. whenever the engineering department feels that the customer requirements were not quite perfectly described
    c. whenever meeting the customer requirements would be too costly
    d. none of the above

20. The goal of Step 1D is to:
    a. determine the percent rework generated by the process with respect to the customer requirement
    b. baseline the process
    c. establish valid and reliable customer requirements
    d. none of the above

# Problems

1. A particular process has a customer requirement of no more than fifty-one. Thirty samples are drawn from a production lot batch and measured, with the results displayed in Table 5–6.

**TABLE 5–6**
Sample Data from Production Lot

| | | | | |
|---|---|---|---|---|
| 50 | 50 | 49 | 50 | 51 |
| 48 | 50 | 50 | 52 | 51 |
| 49 | 50 | 51 | 51 | 49 |
| 50 | 49 | 48 | 50 | 52 |
| 50 | 50 | 49 | 51 | 50 |
| 51 | 50 | 48 | 49 | 52 |

   a. What quantifiable statement can be made about giving the customer the right thing right?
   b. What quantifiable statement can be made about the consistency of giving the customer the right thing right?
   c. Draw the histogram of the sample.
   d. What is the percent scrap associated with this process?

2. For the sample in problem 1, what is the percent scrap of the process if the customer requirement changes to:
   a. no more than fifty
   b. no more than forty-nine
   c. no less than fifty-one
   d. no less than forty-nine

3. Give five examples of a specific and measurable customer requirement.

4. A printing company is concerned that the darkness, or shading, of the print produced on a daily 200-page standard report is not consistent (some pages are noticeably lighter than others and some pages are dark at the top of the page but much lighter on the bottom). The report is printed on one of the older machines and there are concerns that the machine may need to be replaced. Due to the expense associated with the replacement of the machine, the quality control department decides to collect some data to support their claim that the machine needs either an extensive overhaul or replacement. They decide that for the next thirty days they will analyze each report, counting the number of pages that will not pass their quality control check. The data are shown in Table 5–7.

**TABLE 5–7**

Sample Data from Printing Machine

| | | | | |
|---|---|---|---|---|
| 5 | 10 | 7 | 10 | 15 |
| 13 | 10 | 10 | 11 | 9 |
| 10 | 11 | 11 | 8 | 9 |
| 12 | 9 | 10 | 10 | 10 |
| 4 | 10 | 9 | 11 | 16 |
| 7 | 15 | 10 | 5 | 13 |

    a. Draw the histogram.

    b. Compute the quality summary statistics of average, range, and standard deviation.

    c. If the customer requirement is for no more than six pages, what percent scrap will be produced by the process?

    d. Assume that you are going to discuss the results of the study to management. In layman's terms, interpret the results of a, b, and c.

5. For the data in problem 4, compute the scrap for each of the revised customer requirements:

    a. No more than seven pages

    b. No more than eight pages

    c. No more than nine pages

    d. No more than eleven pages

    e. No more than thirteen pages

    f. No more than fifteen pages

    g. No less than fifteen pages

# 6

# Step 2: Flowcharting the Process

## 6.1 Chapter Objectives

After completing this chapter, the student should be able to:

- Understand the concept of flowcharts and the benefits that result when a process is flowcharted, especially in terms of process improvement.
- Learn the basic flowcharting symbols and how to construct a flowchart.
- Know what a theoretical flowchart, actual flowchart, and best flowchart is and the theory behind them.
- Know how to construct theoretical, actual, and best flowcharts and how to use them in process improvement.

## 6.2 Key Terms

Actual Flowchart          Best Flowchart          Theoretical Flowchart

## 6.3 Introduction

Now that the customer requirement has been specifically and measurably identified (Step 1), and initial data regarding the process performance have been collected and analyzed with respect to the requirement, it is time for the second step of the recipe.

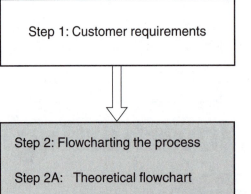

**FIGURE 6–1**
Roadmap Step 2

This step includes looking at what is happening in the process. More specifically, it takes a micro look at the process. As stated in Chapter 4, a vital concept of quality is that complete understanding of a process is the first real step in process improvement. Along with the process checklist, flowchart and flowchart analysis is the mechanism that will not only assist in achieving process improvement, but will serve as the next step in process improvement. Flowcharts provide valuable information and insight into the specific inner workings, or activities, of the process. The roadmap to date is shown in Figure 6–1, and the highlighted box indicates the details of Step 2, which are studied in the chapter.

## 6.4  What Is a Flowchart?

A flowchart is a graphical representation of the specific steps, or activities, of a process. Although flowcharts were used for years in industrial engineering departments in the 1930s, they became very popular in the 1960s when computer programmers used them extensively to map the logic of their various programs. Flowcharts are very powerful tools that:

- Identify inputs and outputs of the process
- Identify the value added and nonvalue added steps
- Provide a common basis of understanding of the process
- Provide a visual representation of event sequence
- Highlight possible indirect paths that may be existing in the process
- Provide invaluable insight when improving a process
- Give an idea as to where bottlenecks may occur
- Give insight into what controls are in place in the process
- Assist in the detection of misunderstanding and miscommunications that may occur with respect to the process
- Give insight into where data collection may occur with relative ease
- Map the logic of the process
- Identify the boundaries of the process
- Provide a mechanism for uncovering problems that may be occurring in the process

As can be seen from the list above, the benefits derived from using flowcharts are numerous and extremely valuable. But the benefits can be realized if, and only if, the flowchart is accurate. This is a very important point to understand. There are times during the construction of a flowchart when people involved in the process may be hesitant to put down exactly what is going on in the process for fear of retribution. Likewise, there are times when assumptions are made that steps occur in a process when in reality those steps may or may not be actually occurring. It must be remembered that as has been stated in previous chapters: garbage in will equal garbage out. Therefore, care must be taken to construct charts that are realistic and possess integrity.

There are many different types of flowcharts. Decision flowcharts, logic flowcharts, systems flowcharts, product flowcharts, and process flowcharts are just a few of the different types of flowcharts that are used in business and government. Even though there are many different types of flowcharts, the steps for flowcharting, shown in Table 6–1, are the same regardless of the type of flowchart.

Because a flowchart is a graphical representation, it is logical that there are symbols to represent the different types of activities. Although there are many different types of flowchart symbols, some of the more common symbols that are used in this text are shown below:

The ellipse represents the beginning or ending (process boundaries) of a process.

The rectangle is used to denote a single step, or activity, in the process.

A diamond is used to denote a decision point in the process. The answer to the question is usually of the yes or no variety.

**TABLE 6–1**

Steps for Constructing a Flowchart

1. Determine the process to be flowcharted.
2. Determine the level of detail to be represented by the flowchart. For example, a macro flowchart would be very global (or higher level) in nature, while a micro flowchart would be very step-by-step (or lower level) oriented.
3. Determine the process boundaries.
4. List the beginning activity of the process.
5. List the sequential activities in the process. (NOTE: A question that can help immensely when laying out a flowchart is "What happens next?")
6. List the ending activity of the process.

  A small circle with either a number or letter inside denotes the point where the process is picked up again.

  An arrow denotes the direction, or flow, of activities in the process.

Within the ellipse (symbol used for process boundaries), the rectangle (symbol used for process activities), and the diamond (symbol used for decision point), a brief description of the step is written. When building a flowchart, it is very important to remember to never sacrifice substance for style. All too many times, the purpose of the flowchart is sacrificed for making a fancy, three-dimensional, multicolored flowchart that will dazzle the audience as opposed to displaying meaningful information about the process.

**EXAMPLE 1**   A local, large volume nursery just received two truckloads of new tomato plants. A flowchart is created not only to spell out the morning and afternoon procedure for thinning out weak plants from lots, but to spell out the daily watering procedure. The flowchart shown in Figure 6–2 is to be used by the rotating new employees who work part time at the nursery.

## **6.5** Roadmap Step 2: Flowcharting the Process

In Step 2, the process is flowcharted to obtain detailed process information and insights as stated earlier. But there is more to flowcharting than just drawing some symbols and arrows and posting them up on the wall. The theoretical, actual, and best

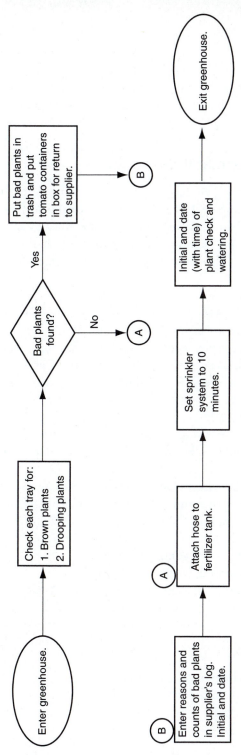

**FIGURE 6–2**
Nursery Flowchart

flowcharts, which are part of Step 2, are flowcharts that not only assist in retrieving as much information as possible about the process, but also combine to give insight into improving the process by involving the people most familiar with the process.

## 6.5.1 Step 2A: The "Theoretical" Flowchart

The recipe steps in flowcharting are relatively straightforward. After the process has been baselined (performed in Step 1 of the recipe), a **theoretical flowchart** is constructed. A theoretical flowchart is a flowchart that is prescribed by some overarching policy, procedure, or operations manual. The theoretical flowchart describes the way the process "should operate." The "should operate" aspect is the Method portion of the *Ms* of a process.

---

**EXAMPLE 1**  The mailroom has an operations manual with specific step-by-step instructions for their various activities (for example, registered mail, certified mail, sorting incoming mail, handling outgoing bulk catalog mailings, and other responsibilities). In Section I of the operations manual, the details for handling incoming mail are spelled out as follows:

1. Take mailroom cart to receiving dock at 7:00 A.M. every weekday morning.
2. Load bags of mail from U.S. Post Office delivery truck to mailroom cart.
3. Return to mailroom.
4. Weigh each bag individually on the mailroom weight scales.
5. Add weight of all bags to determine total weight of incoming mail for that day.
6. Record weight in Master Log.
7. Empty bags on table.
8. Sort First Class bundles from all other bundled mail.
9. Put all other mail on secondary mail table.
10. Unbundle First Class mail and sort in bins for delivery.
11. After all First Class sorting has been completed, check to see if it is your week to deliver mail or your week to sort mail and handle other mailroom duties:

- For mail delivery procedures, turn to Section II.
- For other mailroom duty procedures, turn to Section III.

---

From this section in the operations manual, the theoretical flowchart can now be constructed as shown in Figure 6–3.

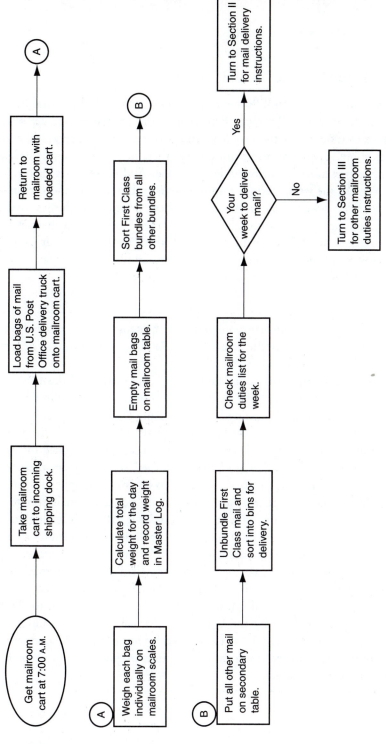

**FIGURE 6–3**
Mailroom Theoretical Flowchart

## 6.5.2 Step 2B: The "Actual" Flowchart

Now that the theoretical flowchart has been completed, the **actual flowchart** is constructed. An actual flowchart depicts how the process is actually working. This is where caution about the flowchart integrity mentioned previously comes into play even more. There are many times when there are better ways of doing things than what is prescribed in an operations manual. And more times than not, people just implement their step-savings improvements to the process and continue business as usual. Even though the process may have been improved with these new steps, people may become very hesitant to mention how the process is actually operating for fear of retribution for not following the prescribed methodology. It is important that integrity exists not only in the actual flowchart, but in all flowcharts as well, especially because these flowcharts are used in process improvement.

---

**EXAMPLE 1**    In continuing with the mailroom example, the department personnel get together and construct a flowchart of the actual steps taken in this procedure. The actual flowchart is shown in Figure 6–4.

When comparing the actual flowchart to the theoretical flowchart it can be seen that the two flowcharts differ. In the actual flowchart, the mailroom employees who pick up the mail each morning have been using the receiving dock scale to weigh the mail. The employees throw the bags on the cart and weigh the entire cart with the bags on it using the loading dock scales. They had previously weighed the cart by itself, so that now when a total weight of "bags and cart" is determined, they can merely subtract the weight of the cart from the total to obtain the weight of the mail. They find this much easier and quicker than taking them back to the mailroom to weigh each of the bags individually. In addition, they found out that the receiving dock scale is calibrated weekly as opposed to the quarterly calibration of the mailroom scale. With this in mind, the employees felt there was no loss in integrity by implementing the change. It is interesting to note that even though they had in fact improved the efficiency of the process (and probably the accuracy of the weight measurement as well) the group was always reluctant to inform management that they were not totally following the procedure as specified by the operations manual. Notwithstanding, the mailroom management and personnel made a sincere commitment to conduct a true process improvement initiative based on the principles of quality, without the fear of retribution or other adverse actions.

---

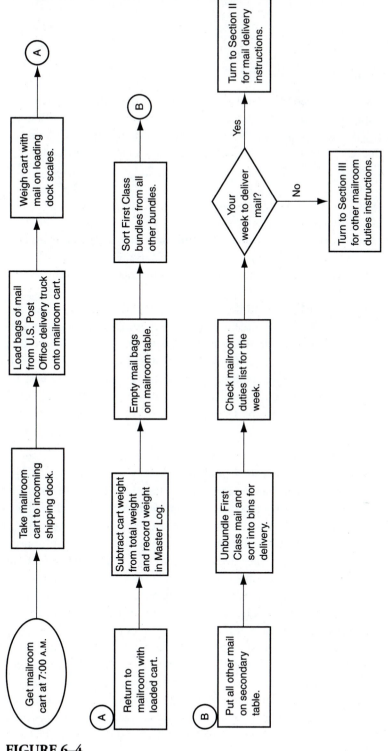

**FIGURE 6–4**
Mailroom Actual Flowchart

175

### 6.5.3 Step 2C: The "Best" Flowchart

Now that the theoretical and actual flowcharts have been completed, there is an observation that needs to be made. The observation is a result of completing Steps 2A and 2B, and sets the stage for effectively conducting Step 2C. Some benefits that will definitely be realized by the people who are constructing the flowcharts include the following:

- They are learning the details of the process.
- They are viewing the process in its entirety.
- They are starting to form a team and use a team approach, where it is "you and me versus the process" as opposed to "you versus me."
- They are in a much better position to recommend improvements to the process.

The last benefit leads to the next recipe step: to construct a **best flowchart**. A best flowchart is where the people involved in the process develop the optimum (or close to optimum) flow based on the process expertise of the group. There are times when the best flowchart is the actual flowchart, and there are other times when the best flowchart is a complete overhaul of the process. At any rate, the best flowchart should reflect improvements that will increase not only the efficiency of the process but also increase quality of the throughput to all of their customers (both internal and external).

---

**EXAMPLE 1**  The mailroom management and workers have now begun to embrace a team process improvement initiative after seeing improvements in several of their processes after flowcharting. Their initiative has the goal of not only increasing the mailroom efficiency, but also increasing the quality of their service to their in-house customers. As a result of their progress, they decided to start with their two biggest customers to determine if there were any improvements in service that the mailroom could offer and possibly accommodate. After that they planned to contact their remaining customers in the same manner. (Note: the group chose this avenue as their approach: starting with a few customers first, and then extending outward.)

Their two biggest customers are the accounts payable and accounts receivable departments. The group decided that the mailroom management would go to these customers, explain the mailroom process improvement initiative, and ask if the current service provided by the mailroom could be improved. The accounts payable department stated that it would help immensely if the numerous bills for the entire company could just be opened, put in a stack, bundled, and delivered. The current manning in accounts payable was not going to increase and because the company was growing by leaps and bounds, so were the bills. The company had always been postured toward a prompt payment belief, and opening the mail was becoming very time consuming and cumbersome, because no one person wanted to be stuck with opening the mail on a daily basis. So the job rotated from employee to employee within the accounts payable department. In essence, the accounts payable management felt that they were beginning to take on additional duties above and beyond ensuring that prompt payments were made to creditors. The mailroom management came back to the group and there was consensus that this was an excellent initial process improvement initiative.

The mailroom group met and developed a proposed best flowchart as shown in Figure 6–5.

The best flowchart differs totally from the theoretical and actual flowcharts. The mailroom, with the accounts payable department as a very vocal supporter, was able to procure a state of the art letter slitter and a paper bundling machine. In addition, the best flowchart also accommodated the weighing of the mail at the receiving dock each morning, as well as several other improvements in delivery schedules and sorting.

The mailroom continued its process improvement initiative with more and more customers, documenting improvements and gathering data to support the increase in efficiency. The benefits, both tangible and intangible, began to be realized:

- The mailroom gained the reputation of being able to do more than just the classic paradigm associated with them of "delivering mail."
- With their improved reputation they soon began to take on more and more responsibilities with their various customers throughout the plants.
- With more responsibilities also came an increased budget with additional manpower.
- The team concept began to grow, along with morale, as they held regular meetings to solve problems.

As one final observation, notice that the implementation of the flowchart recipe step can be conducted on any white collar or blue collar process that delivers products or services to customers. Benefits result with flowcharting and flowchart analysis of any process.

## 6.5.4 Step 2D: Document and Rebaseline the Process

Now that the process has been improved, it is important to rebaseline the process using the techniques taught in Chapter 5. Just because the people involved in modifying the process via flowchart analysis think that the process has been improved does not necessarily mean that it has been. In Chapter 5, the process was baselined when data were

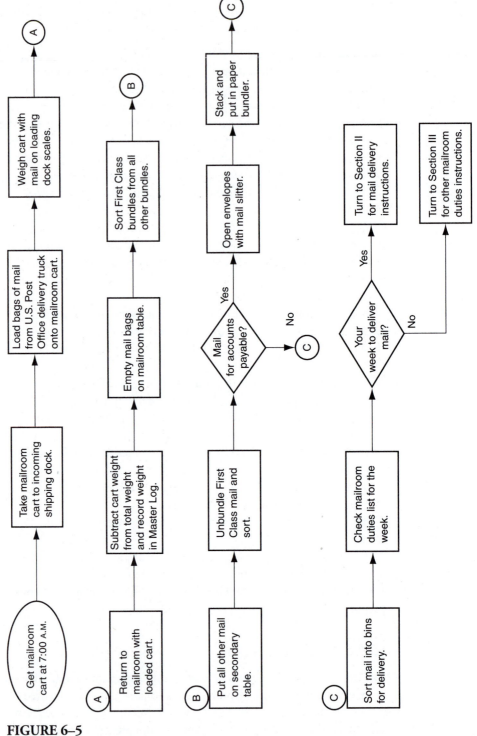

**FIGURE 6–5**
Mailroom Best Flowchart

gathered with respect to the customer requirement, and the mean and range were computed to determine if the process is giving the customer the right thing right the first time. Now that a process change has occurred, rebaselining must occur to determine if the process change improved the quality of the product or service. In addition, this step provides the documentation and audit trail for process improvement.

**EXAMPLE 1**   A delivery company guarantees sixteen-hour delivery of anything within a four-state regional area, or else they provide a refund of 100% plus a 25% inconvenience charge to the customer. As their customer base increases, the management is becoming concerned that their processes may not be optimal because they are starting to pay out more and more refunds to various customers for not delivering within the sixteen-hour time frame. With this in mind, the management decides to initiate a process improvement initiative consisting of drivers, schedulers, and other key company personnel. They gathered sample data on the process, as shown in Table 6–2, before beginning process improvement.

### TABLE 6–2
Delivery Company Sample Statistics (Before Improvement)

| Initial Process Data |
| --- |
| Sample size: 225 deliveries over a 1-week period |
| Mean: 13.5 hours |
| High Value: 18.1 hours |
| Low Value: 9.4 hours |
| Range: 8.7 hours |
| Number of late deliveries: 6 (or 2.67% of the sample size) |

Even though a majority of the deliveries are delivered on time, they decided to conduct a flowchart analysis to see if there is anything within their operations that is causing the inconsistencies. A team was formed to develop the theoretical flowchart, as well as actual and best flowcharts for the six deliveries that were late. They compared the actual flowchart of the late deliveries to the theoretical flowchart and studied the differences. In addition, the group compared all three flowcharts to each other to generate process improvement ideas that would increase the delivery time while not sacrificing safety, damaged packages, or other nonquality characteristics. After briefing management, approval for some selected changes was received. After implementation, the process was once again sampled and data were gathered to see if the changes did indeed improve the process. The process statistics are shown in Table 6–3.

**TABLE 6–3**
Delivery Company Sample Statistics (After Improvement)

| Revised Process Statistics |
| --- |
| Sample size: 238 deliveries over a 1-week period |
| Mean: 12.8 hours |
| High Value: 18.9 hours |
| Low Value: 7.2 hours |
| Range: 11.7 hours |
| Number of later deliveries: 1 (or 0.42% of the sample size) |

Upon further investigation of the high value of 18.9 hours, it was determined that the truck delivering the package was headed to a very remote delivery site and was involved in an accident. Notwithstanding, the new process yielded an overall improvement in the process and was documented and rebaselined accordingly using the new process statistics.

# **6.6** Important Notes about Flowcharts and Flowchart Analysis

Although flowcharts are very valuable tools in process improvement, there are some important notes that need to be made when performing this step of the recipe:

- Before beginning the flowcharting step, care must be taken in the determination of the level of detail (for example, micro versus macro; see Table 6–1). While the general guidelines in flowcharting state that a macro flowchart is for higher level and a micro flowchart is for lower levels, the real determining factor is "What level of detail will yield the most benefit?" Likewise, consistency should be maintained throughout the flowcharting steps in the recipe.
- Caution must be exercised when constructing this step of the recipe. Many times, it is very easy to get carried away and develop a process that overkills, with many unnecessary steps that do not add value to the product or service as it goes through the process. The activities of a process should contribute to the value of the product or service; unnecessary or nonvalue added steps will decrease the efficiency of the process.

## 6.7 Summary

- A flowchart is a graphical representation of the specific steps, or activities, of a process. They are very powerful tools that yield benefits ranging from providing a common basis of understanding to providing an invaluable tool for process improvement.
- Step 2A (Construct the "Theoretical" Flowchart) graphically depicts how the process is intended to operate according to some overarching document or operations policy.
- Step 2B (Construct the "Actual" Flowchart) graphically displays how the process actually operates in reality.
- Step 2C (Construct the "Best" Flowchart) graphically displays the optimum (or close to optimum) process based on the expertise of the group.
- Step 2D (Rebaseline the Process) occurs after implementation of the improvements generated from flowchart analysis. This step is extremely important to validate whether the changes to the process did indeed improve the process. Likewise, this step quantifies the improvement and sets the new baseline.
- Care must be exercised when deciding on which level of flowchart to use (micro or macro). Although macro flowcharts are generally used for higher level and micro flowcharts are generally used for lower levels, the real determining factor is to select the level of flowchart that yields the most beneficial information.
- Care must also be taken to avoid falling into the pitfall of overkill when it comes to developing a best flowchart. For example, it may not be necessary to have 20 process steps just to get a signature on a letter for mailing.

## Review Questions

1. How the process works in reality is shown in a(n):
    a. theoretical flowchart
    b. actual flowchart
    c. best flowchart
    d. none of the above

2. The symbol used to represent the beginning or ending of a process is a(n):
    a. ellipse
    b. rectangle
    c. arrow
    d. diamond
    e. none of the above

3. A flowchart is:
   a. a typed list of activities of a process
   b. only used on blue collar processes
   c. only used on white collar processes
   d. a graphical representation of a product flow
   e. none of the above

4. What is the sequence of steps in the flowcharting stage of the recipe?
   a. Actual flowchart, theoretical flowchart, best flowchart, rebaselining
   b. Theoretical flowchart, best flowchart, actual flowchart, rebaselining
   c. Theoretical flowchart, actual flowchart, best flowchart, rebaselining
   d. Sequence does not matter
   e. None of the above

5. The symbol used for depicting a process activity is a(n):
   a. ellipse
   b. rectangle
   c. arrow
   d. none of the above

6. The benefits of flowcharting include:
   a. highlighting the possible indirect paths that may exist in a process
   b. giving insight into where data collection may occur
   c. mapping the logic of the process
   d. all of the above
   e. none of the above

7. A flowchart that prescribes how a process should operate is a(n):
   a. theoretical flowchart
   b. actual flowchart
   c. best flowchart
   d. a and b
   e. none of the above

8. A pitfall in making a best flowchart is:
   a. not rebaselining the process
   b. overkill by injecting too many unnecessary process steps
   c. ignoring input from a person who is not in the same department as you are
   d. b and c
   e. none of the above

9. Documentation for an audit trail occurs in:
   a. Step 2C: best flowchart
   b. Step 2B: actual flowchart

   c. Step 2A: theoretical flowchart
   d. none of the above

10. The symbol used to denote a decision point is a(n):
    a. ellipse
    b. rectangle
    c. diamond
    d. none of the above

11. A flowchart that depicts a recommended flow is a(n):
    a. best flowchart
    b. actual flowchart
    c. theoretical flowchart
    d. none of the above

12. Flowcharts can be used for:
    a. blue collar processes
    b. white collar processes
    c. product flow
    d. service flow
    e. all of the above

13. A macro flowchart is generally used for:
    a. lower levels
    b. higher levels
    c. both a and b
    d. none of the above

14. The symbol used to denote the point where a process is picked up is a(n):
    a. ellipse
    b. small circle with a letter or number inside
    c. diamond
    d. rectangle
    e. none of the above

15. Rebaselining a process after an improvement:
    a. determines new targets for the improved process
    b. shows the actual flow of activities of the new process
    c. validates whether the improvements implemented did indeed improve the process
    d. assists in determining if the process improvement initiative needs to continue
    e. none of the above

16. The question "What level of detail will yield the most benefit?" assists in determining:
    a. when to use a theoretical flowchart
    b. when to use a best flowchart

c. when to use an actual flowchart

d. whether to use a micro or macro flowchart

e. none of the above

17. A micro flowchart is generally used for:
    a. lower levels
    b. higher levels
    c. both a and b
    d. none of the above

18. The symbol used to denote the flow of activities within a process is a(n):
    a. ellipse
    b. diamond
    c. rectangle
    d. none of the above

19. The question "What happens next?" assists in determining:
    a. what flowchart to use
    b. the ensuing activity in a process
    c. what sample size to use in rebaselining
    d. the confidence level of the statistics
    e. none of the above

20. When constructing a best flowchart, it is wise to:
    a. have process expertise represented on the team
    b. look at each step to determine its value added contribution of the product or service
    c. both a and b
    d. none of the above

## Problems

1. Consider your personal process of leaving work or school for the day, going home, and eating dinner. The process begins when you leave the building (work, school, or another building) and the process ends after the dishes are washed, dried, and put away. With this in mind, and using the symbols discussed in this chapter, construct the actual flowchart for this process.

2. For the process in problem 1, try to improve the process via construction of a best flowchart.

3. Consider a part arriving at a warehouse that needs to be unpacked, logged in as received, and delivered to a specific department within a plant.
   a. Draw a flowchart of how you think the process would actually work.
   b. Is the flowchart you constructed a micro or macro flowchart? How did you determine which type of flowchart it was?

4. Think of a particular process that you are familiar with that you feel may need improvement, preferably at your place of employment. Draw the theoretical flowchart.

5. For the process in problem 4:
   a. Draw the actual flowchart.
   b. Does the actual flowchart deviate from the theoretical flowchart? Why or why not?

6. Continuing with the process in problems 4 and 5:
   a. Draw the best flowchart.
   b. Does the best flowchart deviate from the theoretical flowchart? Why or why not?
   c. Does the best flowchart deviate from the actual flowchart? Why or why not?

7. Give an example of where the product or service would be dangerous in terms of safety (or some other critical parameter) if the actual flow deviated from the theoretical flow.

8. Match the following:

   1. ____  ☐           a. denotes an activity

   2. ____  Ⓐ           b. connects activities

   3. ____  ◇           c. denotes picking up at some other point in the process

   4. ____  →           d. denotes a decision point

9. Answer the following True/False question and explain your reasoning.
   "If the actual and theoretical flowcharts differ, then it is always an indication that the process needs to be improved." True or False

10. a. Briefly describe a process that you are familiar with (other than the ones you selected for the previous problems).
    b. Draw a macro flowchart for the process.

    c. Draw a micro flowchart for the process.

    d. For the process you selected, which flowchart (macro or micro) is better for analysis? Why?

11. Are there times, when looking at process improvement using Step 2, that there would be no need to construct the theoretical flowchart, and the improvement team could immediately construct the actual flowchart and then construct the best flowchart?

    Yes ___ No ___ Explain your answer and cite an example to support your answer.

12. Are there times, when looking at process improvement using Step 2, that there will be no need to construct the actual flowchart, and the improvement team could immediately construct the theoretical flowchart and then construct the best flowchart?

    Yes ___ No ___ Explain your answer and cite an example to support your answer.

13. If it is recommended that the best flowchart is constructed using the expertise of people familiar with the process, why is it necessary to construct the theoretical and actual flowcharts at all?

14. Refer to the nursery flowchart example in Example 1, Section 6.4. What measures do you think are in place, or should be in place, for which rebaselining of the process can occur? Explain your answer.

15. Use the flowchart steps in Table 6–1 to construct a flowchart of the process of coming to class and taking an exam. Be sure to identify all parts of the flowchart steps (for example, process boundaries and micro vs. macro).

# 7

# Step 3: Pareto Analysis and Fishbone Diagram

## 7.1 Chapter Objectives

After completing this chapter, the student should be able to:

- Understand the theory behind the Pareto Principle and be able to apply the concept to processes to identify the largest contributors to process variation or defects.
- Know how to construct a Pareto Diagram and interpret the results.
- Know the theory behind the Ishikawa Diagram, or Fishbone Diagram, and be able to apply the concept to processes to identify sources of variation.
- Know how to construct a Fishbone diagram and interpret the results.
- Understand the importance of rebaselining after implementing initiatives that stem from applying Pareto Analysis and the Fishbone Diagram.
- Understand the concept of subindenture Pareto diagramming and Fishbone diagramming.

## 7.2 Key Terms

| | | |
|---|---|---|
| Cost Pareto | Ishikawa Diagram | Pareto Principle |
| Cause-and-Effect Diagram | Pareto Diagram | Pareto Rule |
| Fishbone Diagram | Pareto Law | |

## 7.3 Introduction

This roadmap step, Pareto analysis and Fishbone diagrams, continues the process improvement initiative. It can be thought of as a transition step in that it not only begins by taking a more microscopic look at the process, but sets the stage for the next roadmap step (control charting). Specifically, Pareto charts and Fishbone diagrams focus on the identification of trouble spots in the process, and it is this identification that serves as a springboard for where in the process to institutionalize control charts. As the student progresses through this chapter, it is important to keep in mind the progression and relationship of the roadmap steps. In the first step, the customer

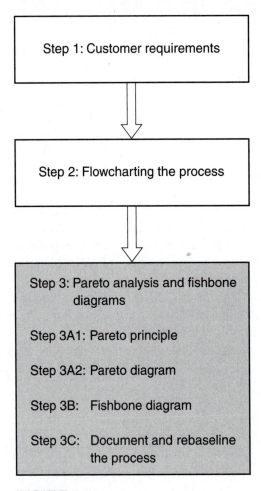

**FIGURE 7–1**
Roadmap Step 3

requirement was obtained and some initial summary statistics were gathered to see how the process performed with respect to the customer requirement. From there, the process was flowcharted to begin initial process improvement. This step of Pareto analysis and Fishbone diagrams continues the roadmap, but with a finer fidelity in terms of process analysis and improvement. The roadmap steps to date, and the stages of Step 3 are shown in Figure 7–1.

## 7.4 Step 3: Pareto Analysis and Fishbone Diagram

Step 3 is the continued microscopic look at the process, only with the focus on the causes of poor performance of the process. It encompasses both the Pareto Analysis and Fishbone Diagram. The Pareto Analysis consist of two main parts: collecting data to apply the Pareto Principle, and construction of the Pareto Chart.

### 7.4.1 Step 3A1: The Pareto Principle

In the nineteenth century, an Italian economist by the name of Vilfredo Pareto determined that 80% of the wealth was in the hands of 20% of the people. His conclusion existed not just in one country but in many countries. Little did the economist know but this basic theory held true for more than just economics and wealth distribution. The **Pareto Principle** (or **Pareto Rule** or **Pareto Law** as it is sometimes called today) is very accurate in most situations in industry and government, as well as most personal life situations. Usually around 80% of the problems or defects experienced in a product or process lie in 20% of the elements or factors. Stated another way: 80% of the effects are caused by 20% of the causes. The Pareto Principle is extremely important because it identifies the largest contributors of process variation that result in poor performance such as defects. Knowing where the biggest contributors lie gives invaluable insight into where to spend precious resources to improve the quality of a product or process.

---

**EXAMPLE 1**   In a small business office, letters are prepared for owner signature. These letters go mainly to various customers, suppliers, and banking institutes. It is mandated by the owner of the company that the letters be professionally written with standard spacing and typing. It is the usual practice that the owner's secretary review the letters before sending them to the owner for signature. The secretary noticed that the number of rewrites of letters before acceptance and submission for signature has increased in the last three months. A decision was made to collect data in an effort to isolate the main causes of rewrites, and then follow up with training in the specific areas that were causing the rework. Data were collected over the ensuing month (when 155 letters were prepared for signature) and arranged in descending order. The collected data are shown in Table 7–1.

**TABLE 7–1**

Checklist Frequency Tally of Causes of Rewrites of Letters

| Cause of Rewrite | Number of Occurrences |
|---|---|
| Misspelled words | 33 |
| Incorrect punctuation | 21 |
| Incorrect spacing | 8 |
| Incorrect signature block | 6 |
| Incorrect heading | 3 |
| | TOTAL = 71 |

Now that the raw data have been collected, a percentage and cumulative percentage can be calculated to identify the largest contributors. The results are shown in Table 7–2.

**TABLE 7–2**

Summary Sheet of Percents and Cumulative Percents of Causes of Rewites of Letters

| Cause | Number of Occurrences | % | Cumulative % |
|---|---|---|---|
| Misspelled words | 33 | 46.5 | 46.5 |
| Incorrect punctuation | 21 | 29.6 | 76.1 |
| Improper spacing | 8 | 11.3 | 87.4 |
| Incorrect signature block | 6 | 8.4 | 95.8 |
| Incorrect heading | 3 | 4.2 | 100.0 |

The first observation from the statistics is that the process does not follow the Pareto split (80/20) exactly. This does not mean that the Pareto Principle does not work. Remember that in our example we only gathered data for one month. Notice, however, that the concept of Pareto generally holds true for this case: 76% of the rework was a result of 40% of the causes, and the two major contributors are mispelled words and incorrect punctuation. The power of Pareto is apparent in this example because training can be developed and aimed at misspelled words and incorrect punctuation, which will hopefully reduce the amount of rework on letters.

Another area that the Pareto law gives insight to is inventory and costs. Consider an airplane for example. There are many parts that go into the construction of an airplane and the parts have different costs. Suppose that the parts can be put into two categories: economic order quantity (EOQ) items and reparable items. EOQ items are usually purchased in bulk (sometimes referred to as bushel basket items) and examples include rivets, screws, and other small parts. Reparable items consist of items such as landing gear, cockpit radios, and large jet engine components, just to name a few. The main distinction between the two categories is that if a part breaks and it is repaired, it is considered a reparable item. Likewise, if a part breaks and it is not repaired, it is considered an EOQ item. As a specific example, a cockpit radio would definitely be repaired when it breaks, and thus is considered a reparable item. A rivet or screw, on the other hand, would not be repaired, and thus is considered an EOQ item. It can likewise be seen that the cost of reparable items is generally much higher than EOQ items. For example, landing gears and cockpit radios cost more than rivets and bolts.

When looking at the parts that go into the construction of an airplane and their associated costs, it is easy to see that Pareto Principle will hold true. There are hundreds of thousands of screws and rivets on an airplane, but only one landing gear. Although a majority of the cost of an airplane lies in the reparable items, a majority of the items on an airplane are EOQ. The Pareto Principle would look very similar to the following (please note that the numbers would not be exactly 80/20, but they would be relatively close):

- EOQ items would constitute approximately 80% of the items on an airplane, and REPARABLE items would constitute approximately 20% of the items.

### ... *BUT* ...

- REPARABLE items account for approximately 80% of the cost and EOQ items account for approximately 20% of the cost.

## 7.4.2 Step 3A2: The Pareto Diagram

A **Pareto Diagram** is a bar chart with the defect categories shown in descending order (from highest category to lowest). The frequency of the categories are shown on the left Y-axis, and the cumulative percent of the categories shown on the right Y-axis. The steps for creating a Pareto diagram are shown in Table 7–3.

**TABLE 7–3**

Steps for Constructing a Pareto Diagram

STEP 1: Determine the time period over which the process will be studied (used for chart header information and documentation).

STEP 2: Determine the categories that will be used for classification. Groups, defects, products, and time required are some examples.

STEP 3: Create a checksheet to tally the frequencies for each of the categories.

STEP 4: After the data collection period, develop a summary sheet using the headers Category (X-axis), Number Defective (left Y-axis), and Cumulative Percent (right Y-axis). Complete the summary sheet using the data from the checksheet (Step 3), putting the categories in descending order.

STEP 5: Draw the bars on the graph in descending order, with the height of each bar corresponding to the left Y-axis value (Number of Defects). Label each category under the appropriate bar on the X-axis.

STEP 6: Using the Cumulative Percent column from the summary sheet, plot a line that corresponds with the right Y-axis. Also note that the right Y-axis should set the 100% value at the same level as the total number of defectives on the left Y-axis.

STEP 7: Put the appropriate header information on the Pareto diagram, inclusive of dates the data were collected. This serves as baseline for improvement.

EXAMPLE 1   In May, a local manufacturing company just installed a new shock absorber line. In order to gather some information about the performance of various numerical control machines, the quality department decides to conduct a Pareto Analysis study. Shock absorbers are spot tested at the end of the assembly line for defects before being packed and prepared for shipment to final assembly divisions. The reject data for three shifts are reported and shown in Table 7–4.

**TABLE 7–4**

Reject Data for Shock Absorber Line

PLANT:   #3-15S
DATE:   June 1

| 1st Shift Production (in units): | 3257 | Number Inspected: | 193 |
| 2nd Shift Production (in units): | 3302 | Number Inspected: | 194 |
| 3rd Shift Production (in units): | 3233 | Number Inspected: | 193 |
| **TOTAL:** | 9792 | **TOTAL Inspected:** | 580 |

**TABLE 7–4** *(cont'd.)*

| Shift | Defect | Number of Defects | Percent of Defects | Percent of TOTAL DEFECTS |
|-------|--------|-------------------|--------------------|--------------------------|
| 1st | Leakers | 6 | 1.03 | 7.14 |
| 2nd | Leakers | 9 | 1.55 | 10.71 |
| 3rd | Leakers | 8 | 1.38 | 9.52 |
| *TOTAL:* | | 23 | 3.97 | 27.38 |
| 1st | Orifice | 2 | 0.34 | 2.34 |
| 2nd | Orifice | 0 | — | — |
| 3rd | Orifice | 6 | 1.03 | 7.14 |
| *TOTAL:* | | 8 | 1.38 | 9.52 |
| 1st | Spot Weld | 14 | 2.41 | 16.67 |
| 2nd | Spot Weld | 12 | 2.07 | 14.29 |
| 3rd | Spot Weld | 15 | 2.59 | 17.86 |
| *TOTAL* | | 41 | 7.07 | 48.80 |
| 1st | Oil (dirt) | 4 | 0.69 | 4.76 |
| 2nd | Oil (dirt) | 1 | 0.17 | 1.19 |
| 3rd | Oil (dirt) | 0 | — | — |
| *TOTAL* | | 5 | 0.86 | 5.95 |
| 1st | Rod (chrome) | 0 | — | — |
| 2nd | Rod (chrome) | 1 | 0.17 | 1.19 |
| 3rd | Rod (chrome) | 0 | — | — |
| *TOTAL* | | 1 | 0.17 | 1.19 |
| 1st | Steel (crimp) | 3 | 0.52 | 3.57 |
| 2nd | Steel (crimp) | 3 | 0.52 | 3.57 |
| 3rd | Steel (crimp) | 0 | — | — |
| *TOTAL* | | 6 | 1.03 | 7.14 |
| **GRAND TOTAL** | | **84** | **14.48** | **99.98** |

From the reject data in Table 7–4, the Pareto statistics are computed and shown in the following list:

- Percent defective:   14.48%
- 76.18% of the defects lie in two (33.33%) of the six failure categories. The two categories are spot welds and leakers.
- Because the spreadsheet was laid out to record the failures by shift, a check can be done to see if there were any shift-unique trends as well:
  – twenty-nine failures occurred on the first shift (34.5%)

  – twenty-six failures occurred on the second shift (31.0%)
  – twenty-nine failures occurred on the third shift (34.5%)

The percentages and cumulative percentages in the summary sheet are then computed and are shown in Table 7–5. The cumulative percentages are needed to determine the 80/20 points.

**TABLE 7–5**
Summary Sheet for Shock Absorber Line

| X-Axis Category | Left Y-Axis (No. defective) | % | Right Y-Axis (Cumulative %) |
| --- | --- | --- | --- |
| Spot Weld | 41 | 48.80 | 48.80 |
| Leakers | 23 | 27.38 | 76.18 |
| Orifice | 8 | 9.52 | 85.70 |
| Steel (crimp) | 6 | 7.14 | 92.84 |
| Oil (dirt) | 5 | 5.95 | 98.79 |
| Rod (chrome) | 1 | 1.19 | 99.98 |

Now that all of the computations associated with the Pareto have been completed, the Pareto diagram is ready to be made. The diagram is shown in Figure 7–2.

The bars represent the defect types arranged in descending order by number of defects per category and are plotted against the left "Y" axis. The cumulative percent of the categories is shown by the X-line plotted on the right "Y1" axis.

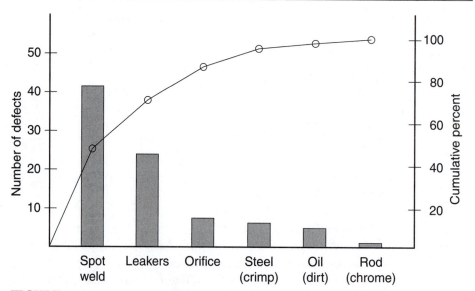

**FIGURE 7–2**
Pareto Diagram for Shock Absorber Defects

Notice that the shock absorber data did follow the Pareto Principle precisely in an 80-20 split (actual split was 76.18%–33.33%). Again, this does not mean that the Pareto Principle is invalid. Recall that it is an approximation. From the summary statistics, process improvement initiatives would be applied in the areas of spot welds and leakers.

The summary sheet for the Pareto serves to point to where process improvement needs to occur and also as a baseline (note that in this case with the new shock absorber line, initial percent defectives are 14.48%). Assume that a process improvement initiative on the spot welds improves the process and the new percent defectives is reduced to 9.24%. Because the process was baselined with the initial Pareto, the percent improvement attributable to the process improvement initiative is 5.24%.

In conducting a Pareto analysis, it is also very feasible and recommended to substitute cost for the frequency of occurrences on the left Y-axis. The cost represents the dollar expenditure associated with downtime, scrap, rework, or other categories. This type of Pareto, a **Cost Pareto**, is very valuable because it details where most of the dollars occur as a result of non-quality.

## 7.4.3 Step 3B: The Fishbone Diagram

Another tool used in analyzing the process, especially with respect to process variability, is the **cause-and-effect diagram**. The cause-and-effect diagram is sometimes referred to as the **Ishikawa diagram**, or the **Fishbone diagram**. Dr. Kaoru Ishikawa of the University of Japan created the cause-and-effect diagram in 1943. At that time he was working with Kawasaki Steel Works demonstrating factor impact and interface on the quality characteristic. Dr. Ishikawa's developed quality technique gained widespread usage throughout Japan and the world. The Fishbone diagram is used for:

• Cause and effect analysis
• Root cause determination of the problem or issue
• Providing a clear graphical display of the sources of variation
• Serving as an excellent tool from which process improvement actions can be originated

To gain an understanding of the use of the Fishbone diagram, consider any process. Recall from previous chapters that after process data are gathered and the histogram is analyzed, the products or services produced by that process will exhibit variability. This variability is attributable to the $M$s of the process (manpower, machinery, materials, methods, and monies). There is variability in the methods (especially if the methods consist predominantly of humans); variability within the machinery (parts within the machine may not be performing optimally due to part wear or other causes); and variability within the material (a roll of steel has property variations throughout the entire roll). These individual variations and combination of individual variations that

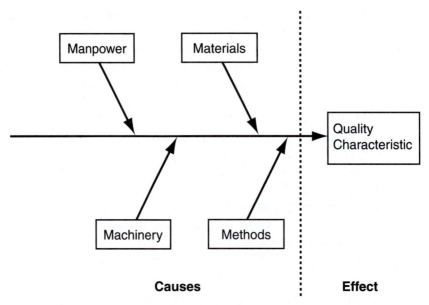

**FIGURE 7–3**
Beginning Design for Fishbone Diagram

add together within the product or service are what the Fishbone diagram attempts to capture. They also are the sources of the variation that can be seen in the Fishbone diagram. The cause-and-effect diagram helps to examine the relationships between the desired effect (or outcome), and the causes (or factors) that influence that effect. Likewise, the Fishbone diagram will assist in the search of the root causes of poor performance. The basic beginning design for the diagram is shown in Figure 7–3.

To construct the Fishbone diagram, the steps shown in Table 7–6 are taken.

**TABLE 7–6**
Steps for Constructing a Fishbone Diagram

STEP 1: Identify what the problem is. The Pareto diagram will help to focus on the major contributing problem area of the process.

STEP 2: Draw the Fishbone diagram skeleton.

STEP 3: Write the effect (or problem) in the quality characteristic box (the head of the fish).

STEP 4: Label the major categories on the "bones" of the fish. These major categories are usually the *M*s of the process and are the overarching major causes of variation. NOTE: other additional major sources of variation categories can be shown on the diagram as well. These additional categories can be identified through brainstorming.

STEP 5: For each bone, brainstorm for possible contributing factors or causes.

STEP 6: After an adequate amount of information has been compiled on the fishbone,

**TABLE 7–6** (*cont'd.*)

evaluate and analyze the possible causes. As a tip for analyzing, causes that occur more than once or in more than one category are very likely candidates.

STEP 7: Take action.

---

**EXAMPLE 1**   In looking at the shock absorber example (Section 7.4.2, Example 1), it is determined from the Pareto analysis that the spot welds will be examined first in an effort to improve the performance of the product. The quality characteristic, or effect, is "spot welds are not holding." The Pareto results indicated that the welds that were not holding were detected in two different places in the plant:

1. After coming off the welding production line where they would break immediately
2. After the finished shocks are put in shipping bins and each bin is spot checked via sampling before shipment

Now that the quality characteristic is defined (Step 1 of Fishbone diagramming), the major categories that contribute to the problem (Step 2 of Fishbone diagramming) are determined to be manpower, machinery, methods, and materials. The initial fishbone is shown in Figure 7–4.

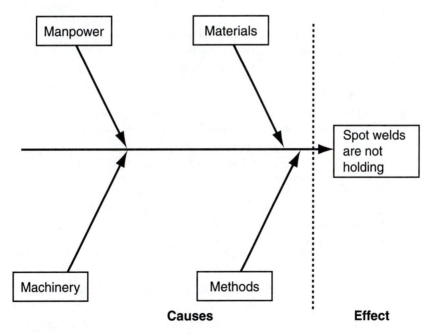

**FIGURE 7–4**
Initial Skeleton of Fishbone Diagram for Shock Absorber Problem

After the initial fishbone skeleton is drawn, the next step in constructing the fishbone diagram is to put the detailed sources under each major category that may be considered the causes.

**EXAMPLE 2**   Some of the detailed causes under each category that have been determined from brainstorming are shown in Table 7–7. Note that some categories have more causes than others but this does not invalidate the diagram or analysis.

**TABLE 7–7**

List of Brainstormed Causes by Category for Spot Welds Not Holding

MACHINERY:
- Welding tip (angle: too much? too little?)
- Age of tip (new? how old?)
- Welding material (composition; purity)
- Welder movement during welding operation
- Age of welders (there are several welders of different ages on the production line)

MANPOWER:
- Experience of union shop welders (excellent experience? little or no experience?)
- Shift (a three-shift operation is employed)
- Training on current welders in the shop

METHODS:
- Approved procedure for spot welding the shocks (inclusive of positional placement of shock absorber tubes, and contact time of welder to shock absorber tube wall)
- Approved procedure for incoming tube dimensions
- Approved procedures for environmental considerations such as cooling and ventilation

MATERIALS:
- Welding material composition
- Shock absorber wall material and composition
- Welding tip material

The identified sources can now be drawn on the fishbone, which are shown in Figure 7–5.

Notice that each item can be decomposed to further contributing root causes (as in MATERIAL and MACHINERY). Although this is not an exhaustively completed

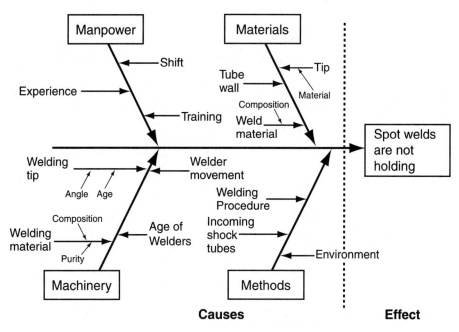

**FIGURE 7–5**
Completed Fishbone Diagram for Shock Absorber Problem

fishbone, the concept of the power of cause identification is certainly demonstrated. The process and what is happening in the process are understood to a very detailed level.

## 7.4.4 Step 3C: Document and Rebaseline the Process

As with flowcharting, it is important once again to rebaseline the process and document the improvements. It should now be evident that every time the process is changed to incorporate an improvement (whether it stems from flowchart analysis, Pareto analysis, or Fishbone diagramming) rebaselining needs to occur. Data should be gathered not only to quantifiably state the impact of the improvement, but to establish a new baseline from which the process now operates. The collected data will again be analyzed using central tendency and variability, as demonstrated in Chapter 5. Again, the measures of central tendency and variability determine if the process is giving the customer a quality product or service.

**EXAMPLE 1**   In continuing with the shock absorber production line example, the per shift initial baseline data are shown in Table 7–8, with the before improvement average (central tendency) and range (variability) inserted for each shift for all defect categories.

**TABLE 7–8**
Initial Reject Data with Baseline Data (Average and Range)

| Shift | Defect | Number of Defects | | |
|-------|--------|-------------------|---|---|
| 1st | Leakers | 6 | | |
| 2nd | Leakers | 9 | | |
| 3rd | Leakers | 8 | | |
| | TOTAL: | 23 | AVG: 7.7 | RANGE: 3 |
| 1st | Orifice | 2 | | |
| 2nd | Orifice | 0 | | |
| 3rd | Orifice | 6 | | |
| | TOTAL: | 8 | AVG: 2.7 | RANGE: 8 |
| 1st | Spot Weld | 14 | | |
| 2nd | Spot Weld | 12 | | |
| 3rd | Spot Weld | 15 | | |
| | TOTAL: | 41 | AVG: 13.7 | RANGE: 3 |
| 1st | Oil (dirt) | 4 | | |
| 2nd | Oil (dirt) | 1 | | |
| 3rd | Oil (dirt) | 0 | | |
| | TOTAL: | 5 | AVG: 1.7 | RANGE: 4 |
| 1st | Rod (chrome) | 0 | | |
| 2nd | Rod (chrome) | 1 | | |
| 3rd | Rod (chrome) | 0 | | |
| | TOTAL: | 1 | AVG: 0.3 | RANGE: 1 |
| 1st | Steel (crimp) | 3 | | |
| 2nd | Steel (crimp) | 3 | | |
| 3rd | Steel (crimp) | 0 | | |
| | TOTAL: | 6 | AVG: 2.0 | RANGE: 3 |

If desired, the data can be rolled up to a higher level for all categories combined, but for purposes of this example, the spot weld, which was identified as the highest contributor from the Pareto will be the focus for initial process improvement. The Fishbone diagram identified several sources of variation and recommended improvements were implemented. Data were again collected on the entire process after the improvement was implemented, and the spot weld data were recorded as shown in Table 7–9.

**TABLE 7–9**
Rebaselined Data for Spot Weld

| Shift | Defect | Number of Defects | | |
|-------|--------|-------------------|---|---|
| 1st | Spot Weld | 7 | | |
| 2nd | Spot Weld | 6 | | |
| 3rd | Spot Weld | 7 | | |
| | *TOTAL:* | 20 | *AVG: 6.7* | *RANGE: 1* |

In looking at the impact of the improvement, it can be seen that the average number of defects went from 13.7 to 6.7 and the variability went from 3 to 1. Not only has process improvement indeed been realized in the reduction of rejects (average), but variability has been reduced as well. The impact of the improvements has been now quantified and a new baseline for the process established.

# 7.5 Extending the Pareto and Fishbone Concepts

In Step 3 of the roadmap, it is recommended to conduct a Pareto analysis and then a Fishbone diagram of the largest contributing factor. After this step has been completed, the process can be further researched by also conducting another Pareto and Fishbone diagram on a subindenture basis as well as a repetitive basis. For example, it may be beneficial for additional process insight to run an initial Pareto, and then a subindenture Pareto of the first bar (or largest contributor) of the initial Pareto chart. It may be beneficial to run a Pareto on one of the bones on the Fishbone diagram. It likewise may be beneficial to construct a Fishbone diagram for all contributing categories that are displayed on a Pareto chart. Although a structured approach of conducting a Pareto and then Fishbone of the largest contributor is recommended, it must be remembered that the goal is process improvement, and the more insight about the process that is obtained, the better the position to recommend improvement areas.

**EXAMPLE 1**   In Section 7.4.1, Example 1, the initial Pareto showed that improper spelling and improper punctuation were the largest contributors. It was decided to Pareto the largest contributor (misspelled words), as shown in Figure 7–6, to get an idea of where to further focus the training.

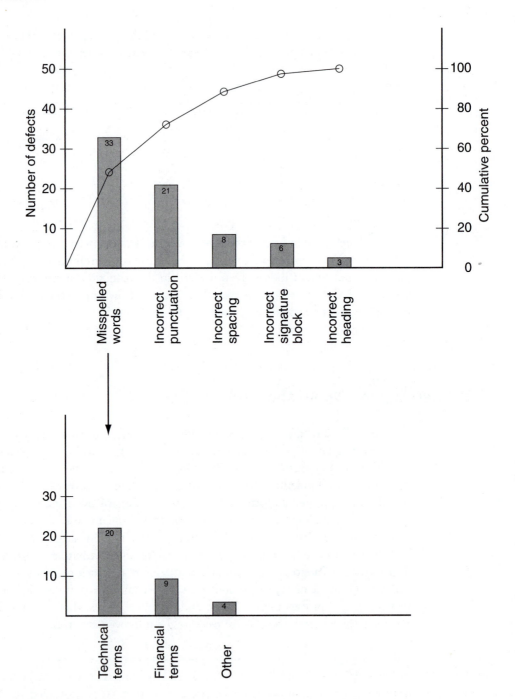

**FIGURE 7–6**
Subindenture Pareto

By conducting this subindenture Pareto on the largest contributor, the training can now be more focused on the specific areas that are causing the greatest problems. A subindenture Pareto was also conducted on the second largest contributor (improper punctuation), but it was soon determined that training on general punctuation was needed and the process improvement initiative on training was modified accordingly.

## 7.6 Summary

- Step 3 involves using the Pareto analysis and Fishbone diagram to focus on specific components of the process.
- The Pareto Principle, which was developed by the Italian economist Vilfredo Pareto, states that 80% of the effects are caused by 20% of the causes. It is also sometimes known as the Pareto rule, or Pareto law.
- The Pareto Principle applies to more than just economics. It is usually true in most processes because approximately 80% of the total variation is caused by 20% of the factors.
- The Pareto diagram is a graphic representation displaying the causes in descending order in terms of the frequency of their occurrences.
- The Fishbone diagram, or Ishikawa diagram (named after its creator Dr. Kaoru Ishikawa), is a technique that gives insight into the causes of product or service variation.
- The head of the Fishbone diagram is the problem (or effect) and the bones are the categories (for example, manpower, machinery, methods, materials, and monies), which are further broken down to identify sources of variation.
- The Pareto and Fishbone diagrams can be also used in a subindenture and repetitive manner.
- Rebaselining of the process needs to occur after process improvement initiatives from the Pareto and Fishbone are implemented as well as anytime in the recipe after an initiative has been implemented.

## Review Questions

1. What is shown on the right Y-axis of the Pareto diagram?
   a. Number of defects
   b. Defect categories
   c. Cumulative percentage of the categories
   d. None of the above

2. What is shown on the head of a Fishbone diagram?
   a. Effect, problem, or quality characteristic
   b. Cause(s) of the effect
   c. One of the M categories
   d. None of the above

3. Which economist established that 80% of the wealth lies with 20% of the population?
   a. Kaoru Ishikawa
   b. Vilfredo Pareto
   c. Dr. Juran
   d. None of the above

4. What is shown on the X-axis of a Pareto diagram?
   a. Number of defects
   b. Defect category
   c. Cumulative percentage of the categories
   d. None of the above

5. The recommended roadmap steps of Step 3 are in order:
   a. Fishbone, Pareto, rebaselining
   b. Pareto, Fishbone, rebaselining
   c. Pareto, rebaselining, Fishbone
   d. none of the above

6. What is shown on the bones of a Fishbone?
   a. Effect, problem, or quality characteristic
   b. Cause(s) of the effect
   c. One of the M categories
   d. None of the above

7. Brainstorming is very effective for the:
   a. left Y-axis of a Pareto
   b. right Y-axis of a Pareto
   c. X-axis categories of the Pareto
   d. none of the above

8. A checksheet is useful for:
   a. tallying the frequency of categories for a Pareto
   b. tallying the number of different categories on a Pareto
   c. both a and b
   d. none of the above

9. A Fishbone diagram is used for:
   a   determining root causes of a problem
   b. cause-and-effect relationships
   c. providing a clear graphical display of sources of variation
   d. all of the above
   e. none of the above

10. What is shown on the left Y-axis of a Pareto diagram?
   a. Number of defects
   b. Defect category
   c. Cumulative percentage of the categories
   d. None of the above

11. A Pareto diagram is most useful when analyzing:
   a. product failures
   b. process failures
   c. service failures
   d. all of the above
   e. none of the above

12. Kaoru Ishikawa developed:
   a. the Fishbone diagram
   b. the Pareto diagram
   c. both a and b
   d. none of the above

13. The *M*s are shown:
   a. as categories on a Pareto chart
   b. on the head of a Fishbone diagram
   c. on the bones of a Fishbone diagram
   d. none of the above

14. A subindenture Pareto is recommended:
   a. when there are many defect categories
   b. when more information is desired
   c. both a and b
   d. none of the above

15. The 80/20 nomenclature is associated with:
   a. Fishbone diagrams
   b. rebaselining
   c. Pareto charts
   d. none of the above

16. A Pareto diagram should show the categories:
    a. in ascending order
    b. in descending order
    c. order does not matter
    d. none of the above

17. Rebaselining in Step 3:
    a. quantifies the results of implemented improvements discovered from using the Pareto and Fishbone
    b. documents audit trails
    c. both a and b
    d. none of the above

18. The Pareto Principle:
    a. is an exact representation of process problems
    b. is an exact representation of product problems
    c. is an exact representation of service problems
    d. none of the above

19. A good tip for determining which cause to start on is:
    a. causes that occur under manpower
    b. causes that occur in more than one category
    c. causes that are brainstormed and voted on as the biggest contributors
    d. none of the above

20. Step 3 in the roadmap:
    a. takes a more microscopic look at the process
    b. can be performed in Steps 1 and 2 of the roadmap
    c. are stand-alone techniques that should never be used together
    d. none of the above

## Problems

1. A company produces shoestrings for two major manufacturers of gym shoes. The shoestrings that are coming out of the manufacturing process are experiencing problems with variability. A process improvement initiative focuses on collecting data to build a Pareto diagram. The data and categories are displayed in Table 7–10.

**TABLE 7–10**

Shoestring Defect Data

| | |
|---|---|
| Number of defects: | 110 |
| Number produced that are too long in length: | 15 |
| Number produced that are too short in length: | 38 |
| Number produced missing both plastic tips on the ends: | 40 |
| Number produced missing one plastic tip on an end: | 12 |
| Other miscellaneous: | 5 |

a. What does the Pareto Principle say with respect to this process?

b. Complete the following table:

| Category | Number of Defects | % | Cumulative % |
|---|---|---|---|
| | | | |

c. Draw the Pareto.

2. Imagine variability existing in a process you use everyday: preparing toast. The variability presents itself in the degree of toasting. Sometimes the toast is burned, sometimes it is not toasted at all (or barely toasted), sometimes it is just right, and sometimes only portions of the bread are toasted. For this situation:

a. Describe the effect.

b. List the "bone" categories and why you feel each category will be needed.

c. Draw the fishbone diagram.

3. Plastic bottles are manufactured for 8 oz, 12 oz, 16 oz, and 24 oz beverage drinks. There have been a growing number of problems with the quality of the bottles. It is decided to track a typical production lot via serial numbering of the lots' bottles. Likewise, a major customer has agreed to put the lot in one of the vending machines in his plant and collect the complaints on the lots in two categories: by bottle size and by defect within the bottle size. The resulting data are shown in Table 7–11.

**TABLE 7–11**
Plastic Bottle Defect Data

| Size | Lot Size | Defect Type | Number of Defects |
|---|---|---|---|
| 8 oz | 500 | Crack(s) in bottle | 3 |
| | | Plastic shavings in bottle | 0 |
| | | Seam split | 7 |
| | | Cap size problems | 2 |
| | | Label does not stay on bottle | 3 |
| 12 oz | 500 | Crack(s) in bottle | 20 |
| | | Plastic shavings in bottle | 24 |
| | | Seam split | 58 |
| | | Cap size problems | 25 |
| | | Label does not stay on bottle | 30 |
| 16 oz | 500 | Crack(s) in bottle | 0 |
| | | Plastic shavings in bottle | 6 |
| | | Seam split | 11 |
| | | Cap size problems | 8 |
| | | Label does not stay on bottle | 7 |
| 24 oz | 500 | Crack(s) in bottle | 4 |
| | | Plastic shavings in bottle | 6 |
| | | Seam split | 7 |
| | | Cap size problems | 9 |
| | | Label does not stay on bottle | 8 |

a. What is the total percent defective?
b. Complete the following table:

| Bottle Size | Quantity Defective | % | Cumulative % |
|---|---|---|---|

c. Draw the Pareto chart.
d. Complete the following statement:

"_____% of the bottle failures occur in _____% of the bottle sizes."

4. Now that the largest defective bottle size has been determined:
   a. Complete the following table for that bottle size:

| Bottle Defect | No. of Defects | % | Cumulative % |
|---------------|----------------|---|--------------|
|               |                |   |              |

   b. Construct the associated Pareto diagram.
   c. Complete the following statement:

   "_____% of the total number of defects in the _____ oz bottle size are caused

   _____% of the defect categories."

5. For the situation described in problem 4, attempt to construct a Fishbone diagram, labeling as many sources of variation as you can. Hint: in this particular bottlemaking company the steps for making plastic bottles are:

   – Heat plastic.
   – Pour plastic into mold pattern for 8 oz, 12 oz, 16 oz, and 24 oz bottles. (Note: mold pattern has dimensions for each critical parameter of bottle height, bottle thickness, bottle shape, screw lid shape, etc.)
   – Cool molds.
   – Remove bottles from molds.
   – Sanitize bottles.
   – Fill bottles with beverage.
   – Install cap on bottle.
   – Affix label to bottle.

# 8

# Control Charts

## 8.1 Chapter Objectives

After completing this chapter, the student should be able to:

- Understand how a statistical process control chart is developed and the theory behind its development.
- Understand the concept of variation as it applies to processes.
- Define the two types of process variation as well as how to identify the types of variation that are resident in a process.
- Build a run chart as well as understand the theory behind the run chart.
- Understand the relationship of run charts to statistical process control charts.
- Understand the theory of combining variation over time (run chart) to develop a control chart.

## 8.2 Key Terms

Common Cause Variation    Run Chart
Control Charts    Special Cause Variation

## 8.3 Introduction

This chapter introduces the concept of control charts, what they are, how they are developed, and what they mean with respect to variation. It is imperative that an understanding of control charts be established before continuing with the next steps of the recipe (variables and attributes control charts). Without a proper understanding, incorrect interpretation will occur and the wrong process improvement action may be taken.

## 8.4 What Is a Control Chart?

As the world continues to strive for a global economy, the marketplace is becoming fiercely competitive. With this competition comes the requirement to deliver products and services to existing customers as well as improving products and services that appeal not only to current customers but to new and potential customers as well. With the production of quality products and services comes the requirement of monitoring processes to ensure that they are indeed producing a quality output. **Control charts**, or statistical process control (SPC) charts, are an excellent technique for accomplishing this requirement. They were first developed by Walter Shewhart in 1924.

In the purest sense, a control chart is a line graph that displays an ongoing picture of what is happening in a process. It is a graphical representation of the variation of a process with respect to time. The process data on control charts are charted in the order of occurrence. Using control charts provides many benefits, some of which include:

- Gaining insight about the process
- Monitoring the process to observe its performance
- Identifying variation and the particular types of variation
- Monitoring and reducing the variation
- Determining the process capability
- Controlling the process
- Identifying the presence of trends, shifts, and other patterns in the process

To understand the concept of statistical process control, it is important to understand the physical construction of the control chart and its relationship to the normal curve. The normal curve is the basis for statistical process control. Recall the basic normal curve from Chapter 3, as shown in Figure 8–1, with the mean as well as the mean ± 3 standard deviations.

Now, the entire distribution rotated 90 degrees counterclockwise and the mean + 3 standard deviations is replaced with the term upper control limit (UCL). Likewise, the mean −3 standard deviations is replaced with the term lower control limit (LCL), as shown in Figure 8–2.

As can be seen from Figure 8–2, the control chart is a graphical derivation of the normal curve.

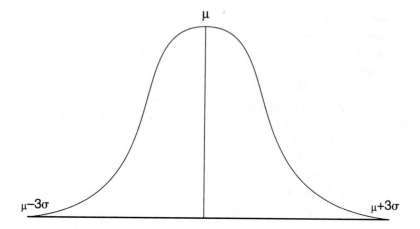

**FIGURE 8–1**
Normal Curve with Mean ± 3 Standard Deviations

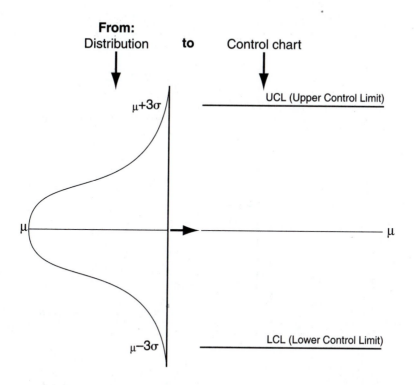

**FIGURE 8–2**
Normal Curve Rotated 90 Degrees to Generate Control Chart

Note: As will be seen under control chart interpretation in Chapters 8 and 9, there is some significance to the values of the mean ± 2 standard deviations. Just as the mean ± 3 standard deviations are called upper and lower control limits, the mean ± 2 standard deviations are called upper and lower warning limits. The interpretation of the warning limits is discussed in the next chapters, but it is only discussed here to show contribution of the mean ± 2 standard deviations to control charts. Figure 8–3 shows the relationship of the normal curve with the mean ± 3 standard deviations and the mean ± 2 standard deviations to the control chart.

As a side note, the inherent capability (which is discussed in detail in Chapter 11) is defined as the area between the mean ± 3 standard deviations.

Now that the physical construction of the control chart is understood, it is time to add the concept of variation to the control chart.

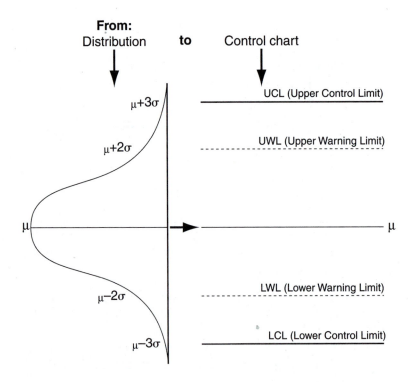

**FIGURE 8–3**
Normal Curve to Control Chart (with Warning Limits)

## 8.5 Types of Variation

When considering a process and the variability that occurs with the process, there are two types of variation that can occur. The two types are **common cause variation** and **special cause variation**. These two types of variation are solely responsible for all of the variation that is exhibited in any process. Each type can be readily identified as long as complete understanding of the theory associated with each type of variation occurs. Furthermore, once complete understanding of common and special cause variation occurs, correct process improvement initiatives can be implemented to the process to reduce these variations that are resident in the process. Once these variations are reduced, then the quality of the product or service delivered to the customer increases.

### 8.5.1 Common Cause Variation

The concept of variation needs to be revisited and understood in complete detail to set the stage for the following discussion. Recall from Chapter 3 that variation is a measure of dispersion and the degree of variation is computed by the statistical measure of variability, more specifically range and standard deviation. Recall also that the normal curve is a pictorial of the process variation as demonstrated in the following example.

---

**EXAMPLE 1**  Management of a local carpentry company that cuts 2 in. × 4 in. treated lumber to 8 ft. lengths for a housing project decides to gather some data to characterize the variation associated with the automated cutting process. The cutting process produces five boards per minute. With this in mind, it is decided to randomly sample ten boards per hour (one every 6 minutes) during the 4-hour cutting period of a typical workday. The rest of the day is spent finishing the wood and delivering the wood to the different customer construction sites. The sampled data (measured in feet) of the lengths are displayed in Table 8–1.

**TABLE 8–1**
Sample Data of Length of Treated 2 in. × 4 in. Lumber

| SAMPLE # | 8:00 A.M. | 9:00 A.M. | 10:00 A.M. | 11:00 A.M. |
|---|---|---|---|---|
| 1 | 8.0 ft. | 8.0 ft. | 8.1 ft. | 8.0 ft. |
| 2 | 8.0 ft. | 7.9 ft. | 8.0 ft. | 7.9 ft. |
| 3 | 8.0 ft. | 8.1 ft. | 7.9 ft. | 8.0 ft. |
| 4 | 8.1 ft. | 8.1 ft. | 7.9 ft. | 8.0 ft. |

*(continued)*

**TABLE 8–1** *(cont'd.)*
Sample Data of Length of Treated 2 in. × 4 in. Lumber

| SAMPLE # | 8:00 A.M. | 9:00 A.M. | 10:00 A.M. | 11:00 A.M. |
|---|---|---|---|---|
| 5 | 8.2 ft. | 7.8 ft. | 8.2 ft. | 7.8 ft. |
| 6 | 8.0 ft. | 8.0 ft. | 8.0 ft. | 8.0 ft. |
| 7 | 8.0 ft. | 8.1 ft. | 8.0 ft. | 8.0 ft. |
| 8 | 7.9 ft. | 8.0 ft. | 8.0 ft. | 8.0 ft. |
| 9 | 8.0 ft. | 7.8 ft. | 8.0 ft. | 8.0 ft. |
| 10 | 8.0 ft. | 8.0 ft. | 8.0 ft. | 8.2 ft. |

The histogram for the sampled data in Table 8–1 is shown in Figure 8–4.

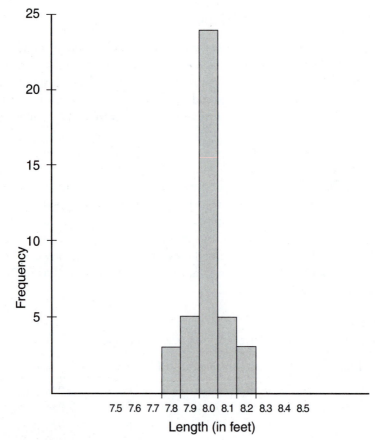

**FIGURE 8–4**
Histogram of Length of 2 in. × 4 in. Wood

As can be seen from the histogram, the process is normally distributed. From the process data, the mean is computed to be 8.0 ft., the standard deviation is 0.09 ft., and the variation can be characterized as follows:

- 68.26% of the process variation is between 7.91 ft. and 8.09 ft.
- 95.45% of the process variation is between 7.82 ft. and 8.18 ft.
- 99.728% of the process variation is between 7.73 ft. and 8.27 ft.

Although the normal curve characterizes variation, there is also another important point that needs to be made. The variation contained in the region mean ± 3 standard deviations is also known as common cause variation. This is the variation that is attributable to the process or system. Common cause variation is produced by the variables of the process and system and their interaction with each other. The common cause variation boundaries of the mean ± 3 standard deviations indicate the boundaries where the process or system operates 99.728% of the time. A pictorial of the region of common cause variation is shown in Figure 8–5.

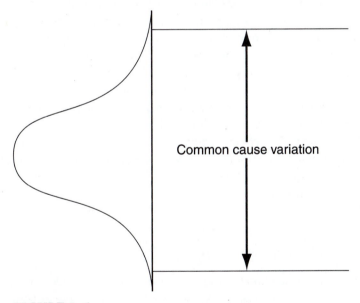

Common cause variation

**FIGURE 8–5**
Normal Curve Identifying Common Cause Variation

EXAMPLE 2    Although there are many examples to explain common cause variation, from writing a particular letter to bending paper clips, this example further clarifies this type of variation. Consider the following process:

- MANPOWER: yourself
- MATERIALS: 0.7 mm pencil; one sheet of 8 ½ in. × 11 in. lined paper; yellow highlighter
- METHOD: write the number 270426 fifteen times with your writing hand; one number per line, the numbers being the height of the line, and a blank line between each number. Write the fifteen numbers in 30 seconds.

Now that the process has been cycled once, look at the numbers. The numbers are not all exactly the same. Some variation exists. The surface smoothness, writer fatigue, number of times the lead broke, and the thickness of the paper, just to name a few, all contribute to the process, or are part of the system. Refocusing on the numbers and all of these factors of the process, use the yellow highlighter to highlight the number(s) that exhibit significant variation from the others. How many did you highlight?

After thinking about the process, it should be obvious that although there is some variation, it is really common cause variation and is brought about by the normal operation of the process. In other words, given the manpower, methods, and materials associated with a process, some common cause variation will be exhibited. This statement is true for every process, not just the preceding number process.

Try it one more time, only this time, change the method slightly to create a new process. Instead of writing the number 270426, write the state where you were born. All other factors ($Ms$) of the process will remain the same, but you will just write the name of your hometown state fifteen times in 30 seconds.

The same phenomenon exists: given the manpower, materials, and methods associated with this new process, variability will be exhibited, but not to any significant degree. And this same phenomenon will exist for every process, from writing a number to producing an automobile, from writing the name of your hometown state to building semiconductors. This example is used in the following section to explain special cause variation as well.

## 8.5.2 Special Cause Variation

The second type of variation (or assignable) that needs to be understood in the world of quality and control charts is special cause variation. Special cause variation is that which is attributable to some special or specific cause not normally associated with the process and system or the interaction of the process and system variables with each other. There is always some special circumstance that is responsible for this type of variation. Likewise, the special cause variation may occur in any part of the process. Unfortunately, many times, when special cause variation occurs, management is all too quick to jump to the conclusion that "the worker" did something wrong. This is not necessarily true, because something special could happen in any of the $Ms$ (manpower included).

EXAMPLE 1    Repeat the portion of Example 2 in Section 8.5.1 that has to do with writing the name of your hometown state fifteen times. This time, however, close your eyes when writing the name of your state for the fifth, sixth, seventh, and eighth times. This could represent a nonplanned temporary power outage, or some other nonplanned contingency. Do not line up your pencil on the paper before closing your eyes. Once you finish writing your state after the fourth time (with your eyes open), close your eyes immediately and write the state with your eyes closed for the next four times. When you have finished, grab the yellow highlighter and highlight the ones that appear to have significant variation outside of what the process called for (for example, height of the lines on the paper). Notice anything? The ones highlighted should correspond to the times when your eyes were closed. The event of closing your eyes was not designed into the process. It is not part of the normal workings of the process. It represented some special circumstance.

In looking at special cause variation and the normal curve, it can be seen that this variation does not occur within the normal boundaries of the process (mean ± 3 standard deviations). Rather, this variation lies outside of these boundaries as shown in Figure 8–6.

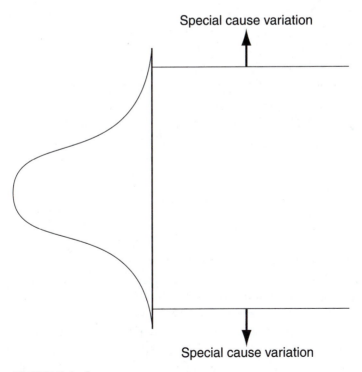

**FIGURE 8–6**
Normal Curve Identifying Special Cause Variation

**EXAMPLE 2**   In revisiting the carpentry example (Section 8.5.1, Example 1), suppose that the data collected are shown in Table 8–2.

**TABLE 8–2**
New Sample Data of Length of Treated 2 in. × 4 in. Lumber

| SAMPLE # | 8:00 A.M. | 9:00 A.M. | 10:00 A.M. | 11:00 A.M. |
|---|---|---|---|---|
| 1 | 8.0 ft. | 8.0 ft. | 8.1 ft. | 8.0 ft. |
| 2 | 8.0 ft. | 7.9 ft. | 8.0 ft. | 7.9 ft. |
| 3 | 8.0 ft. | 8.1 ft. | 7.9 ft. | 8.0 ft. |
| 4 | 8.1 ft. | 8.1 ft. | 7.9 ft. | 8.0 ft. |
| 5 | 8.2 ft. | 7.8 ft. | 8.2 ft. | 7.8 ft. |
| 6 | 8.0 ft. | 8.0 ft. | 18.0 ft. | 8.0 ft. |
| 7 | 8.0 ft. | 8.1 ft. | 8.0 ft. | 8.0 ft. |
| 8 | 7.9 ft. | 8.0 ft. | 8.0 ft. | 8.0 ft. |
| 9 | 8.0 ft. | 7.8 ft. | 8.0 ft. | 8.0 ft. |
| 10 | 8.0 ft. | 8.0 ft. | 8.0 ft. | 8.2 ft. |

The process can now be characterized as follows:

- The process average is 8.25 ft.
- The standard deviation is 1.58 ft.
- Common cause variation will be between 3.51 ft. (mean − 3 standard deviations) and 12.99 ft. (mean + 3 standard deviations).
- Special cause variation will be below 3.51 ft. and above 12.99 ft.
- The 2 in. × 4 in. piece of wood that was cut to 18.0 ft. can definitely be attributed to some special cause.

# 8.6 Run Charts

Now that the linkage of variation to the control charts has been established, the linkage of time of occurrence needs to be established. To do this, a **run chart** will be used. A run chart is a graphical display of data in the order they occur. It is used to help determine the presence of special cause variation as well as help determine trends and shifts over the specified period of time. The steps in the construction of a run chart are shown in Table 8–3.

**TABLE 8–3**
Steps for Constructing a Run Chart

| STEP 1: | Determine the process variable of interest. |
|---|---|
| STEP 2: | Draw the vertical axis (Y-axis). |
| STEP 3: | Label the Y-axis with the variable name. |
| STEP 4: | Determine the scale for the vertical axis and mark the axis interval accordingly. |
| STEP 5: | Draw the horizontal axis (X-axis). |
| STEP 6: | Label the X-axis with the unit of time in which the process variable data will be collected. |
| STEP 7: | Determine the scale for the horizontal axis and mark the axis intervals accordingly. |
| STEP 8: | Plot the data values on the graph one by one, being extremely careful to maintain the order in which the numbers were collected. |
| STEP 9: | Connect the points. |
| STEP 10: | Find the central tendency measure of average (or mean) and draw it on the graph as a centerline. |

**EXAMPLE 1**   A manufacturing process produces the movement mechanism for grandfather clocks. One of the gear wheels has a center hole in which a shaft is to be fitted. The engineering blueprint calls for the hole to have a diameter of 0.255 in. The gear wheel is a very high precision part, and there is concern that the tooling on the machinery may need to be changed more often than the once per two-week replacement interval that is currently being adhered to. Furthermore, the current tooling replacement policy calls for the tooling to be changed every other Saturday and the machine runs five days a week. It is decided that a run chart will be helpful in determining if there are any negative trends that are developing with the current tooling change policy that is in existence. It is further decided to gather six weeks of data, which will include three tooling change periods. One gear wheel will be sampled each day for measurement, making the total sample size thirty. The collected data are shown in Table 8–4. The run chart for the gear wheel is shown in Figure 8–7.

**TABLE 8–4**
Sample Data of Gear Wheel Center Hole Dimension

|  | Monday | Tuesday | Wednesday | Thursday | Friday |
|---|---|---|---|---|---|
| **Week 1** | 0.255 in. | 0.255 in. | 0.254 in. | 0.256 in. | 0.256 in. |
| **Week 2** | 0.257 in. | 0.256 in. | 0.259 in. | 0.259 in. | 0.260 in. |
| **Week 3** | 0.255 in. | 0.255 in. | 0.255 in. | 0.256 in. | 0.255 in. |
| **Week 4** | 0.256 in. | 0.256 in. | 0.256 in. | 0.258 in. | 0.258 in. |
| **Week 5** | 0.255 in. | 0.254 in. | 0.255 in. | 0.255 in. | 0.255 in. |
| **Week 6** | 0.255 in. | 0.256 in. | 0.258 in. | 0.259 in. | 0.259 in. |

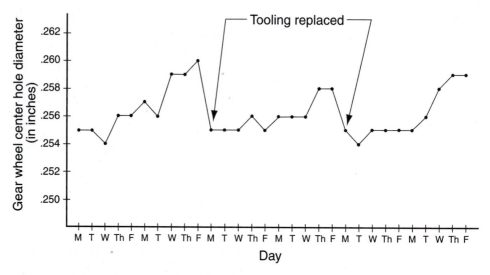

**FIGURE 8–7**
Run Chart of Gear Wheel Center Hole Diameter

Notice that the run chart helped identify a trend that appears to occur every two weeks: the tooling seems to degenerate rapidly toward the last couple of days before replacement, which is reflected by the larger center hole diameter. This trend tells the analysts that the tooling replacement policy of every two weeks may need to be revisited and possibly updated.

Example 1 also points out a very important caution that needs to be addressed when considering run charts. There is a tendency to view every single variation in the run chart as being of extreme importance, and this may not be necessarily true. The goal of the run chart is not to focus on every single movement in variation, but rather to focus on the important changes in the process that are visible on the run chart.

## 8.7 Putting Variation and Time Together

In developing the theory behind the control chart, the control chart is a combination of:

- The normal distribution being rotated 90 degrees counterclockwise
- The theory of common and special cause variation explored as it relates to the normal curve
- The concept of time via the run chart

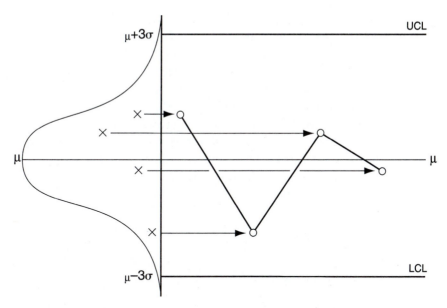

**FIGURE 8–8**
Relationship of Normal Curve Data Points to Control Chart Data Points

Keeping in mind these three points, the theory of the control chart can now be understood. The control chart graphically displays the variation of a product or service over time. The control chart, as seen in the following chapters, is used exhaustively to improve the process by identifying trends, shifts, or out of control points that are resident in the process. Figure 8–8 depicts the relationship of the normal curve to the control chart.

## 8.8 Summary

- Control charts were first developed by Walter A. Shewhart in 1931.
- A control chart is a graphical representation of the variability of a process with respect to time.
- A control chart provides many benefits, some of which include the identification of variation within a process, identification of the types of variation, determination of inherent capability of a process, and monitoring the process to observe its performance as well as controlling it.
- The control chart is generated by taking the mathematical concept of a normal distribution and rotating it 90 degrees counterclockwise. The upper control limit (UCL) is the mean + 3 standard deviations, the lower control limit (LCL) is the

mean – 3 standard deviations, and the centerline of the control chart is the mean (or average) of the normal curve.

- Inherent capability, which is discussed in detail in Chapter 11 is defined as the area between the mean ± 3 standard deviations.
- Common cause variation is variation that is naturally attributable to the process or system. On a control chart, common cause variation occurs between the mean + 3 standard deviations and the mean – 3 standard deviations.
- Special cause variation is variation that is attributable to some special or specific cause not associated with the process and system or the interaction of the process and system variables with each other. On a control chart, common cause variation occurs above the mean + 3 standard deviations or below the mean – 3 standard deviations.
- A run chart is a graphical display of the process data in the order they occur, or with respect to time.
- The combination of the concept of variation and run charts helps explain the theory behind the construction of a control chart.

## Review Questions

1. The warning limits in a control chart occur at the:
   a. mean ± 1 standard deviation
   b. mean ± 2 standard deviations
   c. mean ± 3 standard deviations
   d. none of the above

2. A graphical display of the process data in the order they occur is called:
   a. a run chart
   b. the area where common cause variation occurs
   c. the area where special cause variation occurs
   d. none of the above

3. Variation that is attributable to the process and system is called:
   a. special cause variation
   b. common cause variation
   c. both a and b
   d. none of the above

4. The variation that exists within a process is graphically shown on:
   a. a run chart
   b. a normal curve

   c. a process checklist
   d. a and b
   e. none of the above

5. Common cause variation will occur:
   a. outside of the upper and lower control limits
   b. outside the upper and lower warning limits
   c. inside the upper and lower control limits
   d. none of the above

6. Inherent capability is shown on the control chart as the area:
   a. between the mean ± 1 standard deviation
   b. between the mean ± 2 standard deviations
   c. between the mean ± 3 standard deviations
   d. none of the above

7. Variation that is attributable to causes that do not exist naturally within the process or system is called:
   a. special cause variation
   b. common cause variation
   c. both a and b
   d. none of the above

8. Some benefits of control charts include:
   a. controlling the process
   b. identifying the patterns that exist in the process
   c. monitoring process variation
   d. all of the above
   e. none of the above

9. In looking at the physical construction of a control chart, it can be seen that the normal curve is:
   a. rotated 90 degrees clockwise
   b. rotated 90 degrees counterclockwise
   c. rotated 180 degrees clockwise
   d. rotated 180 degrees counterclockwise
   e. none of the above

10. Control charts were first developed by:
   a. Walter Shewhart in 1931
   b. Dr. Deming in 1931
   c. Dr. Juran in 1931
   d. none of the above

11. Which of the following statements is true?
    a. Control limits occur at the mean ± 2 standard deviations.
    b. Warning limits occur at the mean ± 3 standard deviations.
    c. Both a and b.
    d. None of the above.

12. A completely automated process is set up to run at the prescribed speed. Back-orders are starting to build up, so it is decided to triple the speed of the machinery for a period of three hours in order to catch up. This example would probably result in:
    a. common cause variation
    b. special cause variation
    c. both a and b
    d. none of the above

13. Common cause variation is represented by:
    a. 68.26% of the variation
    b. 95.45% of the variation
    c. 99.728% of the variation
    d. none of the above

14. Special cause variation is represented by:
    a. 68.26% of the variation
    b. 95.45% of the variation
    c. 99.728% of the variation
    d. none of the above

15. Common cause variation is caused by:
    a. the $M$s of a process
    b. monitoring a process with a run chart
    c. monitoring a process with a control chart
    d. none of the above

## Problems

1. Select a process that you are familiar with and:
   a. Describe the process.
   b. List the $M$s of the process.
   c. Describe a situation where special cause variation would occur.

2. A process yields the following data:

| TIME | Monday | Tuesday | Wednesday | Thursday | Friday |
|---|---|---|---|---|---|
| 7 A.M. | 72 | 75 | 74 | 72 | 70 |
| 9 A.M. | 73 | 76 | 72 | 71 | 78 |
| 11 A.M. | 72 | 77 | 70 | 75 | 74 |
| 1 P.M. | 74 | 74 | 98 | 73 | 75 |
| 3 P.M. | 73 | 72 | 73 | 72 | 76 |
| 5 P.M. | 74 | 73 | 76 | 74 | 72 |

    a. What is the process mean?
    b. What is the process standard deviation?
    c. What is the upper control limit of the process?
    d. What is the lower control limit of the process?
    e. What is the upper warning limit of the process?
    f. What is the lower warning limit of the process?

3. For the process in problem 2, complete the following statements:
    a. Common cause variation will occur at values between ____ and ____.
    b. Special cause variation will occur when the values are below ____ or above ____.
    c. Are there any special cause variation data points that occurred in the process? If so, list them.

4. For the process in problem 2, construct the run chart.

5. Incoming telephone calls to a 24-hour catalog ordering department are recorded per hour as follows:

| | |
|---|---|
| 12 midnight to 1 A.M.: | 18 |
| 1 A.M. to 2 A.M.: | 16 |
| 2 A.M. to 3 A.M.: | 15 |
| 3 A.M. to 4 A.M.: | 22 |
| 4 A.M. to 5 A.M.: | 25 |
| 5 A.M. to 6 A.M.: | 17 |
| 6 A.M. to 7 A.M.: | 14 |
| 7 A.M. to 8 A.M.: | 13 |
| 8 A.M. to 9 A.M.: | 24 |

| | |
|---|---|
| 9 A.M. to 10 A.M.: | 27 |
| 10 A.M. to 11 A.M.: | 33 |
| 11 A.M. to 12 noon: | 25 |
| 12 noon to 1 P.M.: | 19 |
| 1 P.M. to 2 P.M.: | 16 |
| 2 P.M. to 3 P.M.: | 22 |
| 3 P.M. to 4 P.M.: | 20 |
| 4 P.M. to 5 P.M.: | 21 |
| 5 P.M. to 6 P.M.: | 18 |
| 6 P.M. to 7 P.M.: | 11 |
| 7 P.M. to 8 P.M. | 19 |
| 8 P.M. to 9 P.M.: | 21 |
| 9 P.M. to 10 P.M.: | 14 |
| 10 P.M. to 11 P.M.: | 16 |
| 11 P.M. to 12 midnight: | 13 |

a. What is the average number of incoming telephone calls received per hour?
b. What is the standard deviation?
c. What are the upper and lower control limits?
d. What are the upper and lower warning limits?
e. Are there any data points above the upper control limit or below the lower control limits? If so, list them.
f. Construct a run chart.

6. Think of two processes that you are familiar with that will periodically exhibit special cause variation. For each process:
a. Describe the process.
b. Describe the circumstances that led to the process exhibiting special cause variation.
c. Estimate the mean of the process.
d. Estimate the value of the special cause variation.
e. From the presence of special cause variation, what changes, if any, would you implement to the process?

7. Think of two processes that you are familiar with that will seldom exhibit special cause variation. For each process:
   a. Describe the process.
   b. Explain why the process will seldom exhibit special cause variation.
   c. Estimate the mean of the process.

8. Think of three processes (other than the ones selected in any of the previous chapter questions) in which conducting a run chart would be beneficial. For each process:
   a. Describe the process.
   b. Explain why using a run chart would be beneficial.
   c. Explain what process insight you think will be gained by using a run chart.

9. It is felt that a repetitive service process is experiencing intermittent delays and no one in the department is able to generate any specific recommendations other than some general hypotheses on "why" the delays might occur. It is decided to gather some sequential data and construct a run chart. It is felt that the run chart can give some insight as to where, when, and why the delays are occurring. The processing time for the transactions (in minutes) are recorded in order of occurrence as shown in Table 8–7.

**TABLE 8–7**
Sample Data of Transaction Processing Times

| | | | |
|---|---|---|---|
| 35 | 40 | 36 | 34 |
| 38 | 37 | 41 | 34 |
| 36 | 35 | 39 | 33 |
| 33 | 35 | 32 | 35 |
| 32 | 36 | 35 | 34 |

   a. What is the process mean?
   b. What is the standard deviation?
   c. Construct a run chart.
   d. Is there special cause variation in the process?
   e. How can you explain the hypothesis that the process is experiencing delays?

10. In problem 9, the process improvement team now felt they had enough information about the process, proceeded to brainstorm some ideas, and decided to implement some changes to the process to try to remove the delays. The group implemented all of the changes at the same time without exhaustively conducting the Deming Plan/Do/Check/Act cycle. The resulting data (in minutes) are shown in Table 8–8.

**TABLE 8–8**

Sample Data of Transaction Processing Times after Implementing Improvements

| | | | |
|----|----|----|----|
| 36 | 35 | 37 | 38 |
| 36 | 41 | 40 | 38 |
| 39 | 39 | 38 | 37 |
| 35 | 35 | 35 | 33 |
| 36 | 35 | 36 | 37 |

a. What is the process mean?

b. What is the standard deviation?

c. Construct a run chart.

d. Is there any special cause variation in the process?

e. When comparing the results of this problem to the results of problem 10, what can be said about the old process versus the new process?

f. What could the process improvement group have done differently, if anything?

# 9

# Step 4A: Variables Control Charts

## 9.1 Chapter Objectives

After completing this chapter, the student should be able to:

- Understand the concepts of variables control charts.
- Understand what variables control charts are used for.
- Identify the four basic types of variables control charts and what type of situation calls for which type of control chart.
- Know how to construct the four different types of variables control charts.
- Know how to distinguish between which control chart factors to use for the different charts.
- Understand how to interpret the control charts and identify various patterns that may exist on the control charts.

## 9.2 Key Terms

Average and Range Chart
Average and Sigma Chart
Control Chart Factors

Individuals and Range
   Chart
Median and Range Chart

## 9.3 Introduction

Now that the concept behind the meaning and construction of the control chart has been established, the next steps of the recipe, variables and attributes control charts, can be conducted. Control charts play a vital role in process improvement, and companies in the automotive industry and semiconductor industry, just to name a few, have implemented statistical control charts with great success. Control charts are

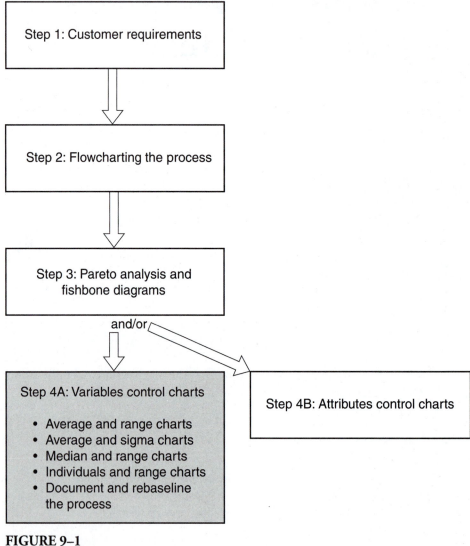

**FIGURE 9–1**
Roadmap Step 4A

based on statistics as a foundation. Recall from Chapter 8 the concepts of run chart, assignable cause, and common cause variation are all based on statistical foundations. Control charts point to specific areas in the process that are causing problems in variation, which in turn, result in poor quality products and services. Variables control charts are explored in this chapter and attributes control charts are studied in Chapter 10. Both types of charts are presented in an applications manner. Both Chapters 9 and 10 are part of Step 4 of the roadmap, because there will be times when variables control charts are used, times when attributes control charts are used, and times when both types of control charts are used. Figure 9–1 depicts the roadmap to date and highlights the variables chart portion of Step 4.

## 9.4  Variables Control Charts Overview

Before jumping in and making a variables control chart, there are certain fundamental concepts about the charts and subgroup sizes that need to be understood. It is this understanding that sets the stage for correct selection of which control chart to use, as well as how to construct the chart.

### 9.4.1 Objectives of Variables Control Charts

There are many reasons for using variables control charts, some of which include:

- To identify critical problem areas, or sources of variation, in which remedial action is needed
- To identify when the process should be left alone and no action taken
- To determine the process capability
- To assist in the establishment of specifications, or to assist in determining when specifications need to be changed
- To monitor the process
- To gather information that is used in inspection or acceptance procedure decisions
- To help understand the process and make personnel familiar with the theory and usage of control charts
- To assist in the determination of the reliability of the product or service

### 9.4.2 Types of Variables Control Charts

Recall from Chapter 2 that variables data have to do with measurements. Measurement of length and measurement of weight are some examples. There are several types of variables control charts and each chart has its own purpose. The basic types of variables control charts that are addressed in this chapter are:

- **Average and Range chart**
- **Average and Sigma chart**
- **Median and Range chart**
- **Individuals and Range chart**

Each chart is discussed in a different section in this chapter and explained in terms of its purpose and application. The order of presentation of the charts in this chapter coincides with their popularity in terms of application.

## 9.4.3 Subgroup and Subgroup Size Considerations

There are a couple of observations that need to be made with regard to subgroups and subgroup sizes for variables control charts. Walter Shewhart developed and introduced the concept of subgroups for control chart purposes. He suggested typically four to five items per subgroup, which allows for within subgroup variation. Likewise, he recommended that enough subgroups be collected so that trends, level shifts, and patterns could be recognized.

The subgroup size should not be arbitrary because it is the average and range of a subgroup that represent a plotted data point on the control chart. As an example, subgroups may represent some period of time or a group of incoming lots. The subgroups should be similar (for example subgroups collected at 8:00 A.M. each morning) but

**TABLE 9–1**

Steps in Constructing a Variables Control Chart

| |
|---|
| STEP 1: Select the process to be studied. |
| STEP 2: Determine which variables chart will best fit the process. |
| STEP 3: Determine the subgroup size and frequency, as well as the number of subgroups to be collected. |
| STEP 4: Collect the data on a data collection sheet. |
| STEP 5: Determine the mean and range for each subgroup. |
| STEP 6: Calculate the grand average or the average of all of the individual subgroups. The grand average is the centerline of the X-bar chart. |
| STEP 7: Calculate the grand range, or the average of all of the individual subgroup ranges. The grand range is the centerline of the R-chart. |
| STEP 8: Compute the control limits for the charts(s). |
| STEP 9: Graph the chart(s). |

Note: What is calculated in Steps 5–7 depends on which type of variables chart is selected. In the steps shown in Table 9–1, the mean and range for an X-bar and R-chart were selected and used for demonstration purposes.

allow for variation to exist between the subgroups. The variation between subgroups can sometimes be thought of as being a function of subgroup frequency.

The best way to sum up subgroup considerations is that it is desirable to have a subgroup size that provides the opportunity for both within group variation and between group variation.

### 9.4.4 Steps in Making a Variables Control Chart

The construction steps are basically the same for all of the variables control charts, except that the statistics computed from the subgroups will vary depending on whether an X-bar and R-chart, X-bar and Sigma chart, Median and Range chart, or Individuals and Range chart is being used. The basic steps in control chart construction are shown in Table 9–1.

## 9.5  Step 4A: Variables Control Charts

Now that an understanding of the fundamentals is achieved, the four variables control charts can be discussed and applications of their usage exhibited. Remember that for each of the variables control charts, variables data are used for each type of chart. Likewise, it is only the mathematical measure of average, median, or sigma that changes when considering the difference between the variables control charts.

### 9.5.1 Average and Range Charts

Average and Range charts are usually used together as a pair of charts and are most often recognized and referred to as the X-bar and R-chart. These charts are definitely one of the most powerful types of control charts. The X-bar (or average) chart is a plot of subgroup averages (central tendency). The R-chart is a plot of the subgroup ranges (or variability). Recalling that quality is defined by the two mathematical measures of central tendency and variability, it can be easily seen why the X-bar and R-charts are instrumental in process quality.

The following example not only demonstrates the construction of an X-bar and R-chart, but introduces **control chart factors** as well. Control chart factors are factors that are used in the calculation of the upper and lower control limits for different control charts. They are handy and easily used after the data have been collected and the average and range computed. There are different control chart factors, depending on which control chart is used:

- $A_2$ is the control chart factor used for X-bar charts.
- $D_3$ and $D_4$ are the control chart factors used for R-charts.
- $B_3$ and $B_4$ are the control chart factors used for Sigma charts.

- $E_2$ is the control chart factor used for Individuals charts.
- $A_6$ is the control chart factor used for Median charts.

<hr>

**EXAMPLE 1**  A small electrical company has an automated process that the quality team thinks needs to be monitored. The circuit breakers are one of the company's prime products and the market potential for growth appears to be quite sizable. The company is beginning to experience huge increases in sales from housing and office building contractors both in the local community as well as out of state contractors.

The parameter of interest is the width (2.0 in.) of the circuit breakers. Too wide a circuit breaker will result in the breaker not fitting in the fuse box. Likewise, too narrow of width will cause play once the breaker has been installed, which could result in a dangerous situation. It is imperative that the process producing the circuit breakers be stable. Because the dimension of interest is width, a variables control chart is in order and the quality team determines that the X-bar and R-chart will be of most benefit. It is further decided that because the circuit breakers are produced at the rate of twenty per hour and the process runs one eight-hour shift per day for five days, subgroups will consist of five breakers. For this study the group decides to draw one subgroup twice per shift for five consecutive days, which will result in ten subgroups. The raw data collected for this study are shown in Table 9–2.

**TABLE 9–2**
Raw Data for Circuit Breaker Problem

| Sample No. | Monday A.M. | Monday P.M. | Tuesday A.M. | Tuesday P.M. | Wednesday A.M. | Wednesday P.M. | Thursday A.M. | Thursday P.M. | Friday A.M. | Friday P.M. |
|---|---|---|---|---|---|---|---|---|---|---|
| 1 | 2.0 | 2.0 | 2.0 | 2.1 | 2.1 | 2.0 | 1.9 | 1.9 | 2.0 | 2.0 |
| 2 | 2.0 | 2.1 | 2.1 | 1.9 | 1.9 | 1.9 | 1.9 | 2.0 | 2.0 | 2.0 |
| 3 | 2.1 | 2.1 | 2.1 | 3.2 | 2.1 | 2.0 | 2.1 | 2.0 | 2.1 | 2.0 |
| 4 | 1.9 | 1.9 | 2.0 | 3.7 | 2.0 | 2.0 | 2.0 | 2.0 | 2.0 | 2.0 |
| 5 | 2.0 | 2.0 | 2.1 | 3.6 | 2.0 | 2.1 | 1.9 | 1.0 | 1.9 | 2.1 |

Once the raw data have been collected, the mean and range for each of the subgroups can be determined as shown in Table 9–3.

**TABLE 9–3**
Circuit Breaker Data with Mean and Range of Subgroups

| | Monday | | Tuesday | | Wednesday | | Thursday | | Friday | |
|---|---|---|---|---|---|---|---|---|---|---|
| Sample No. | A.M. | P.M. | A.M. | P.M. | A.M. | P.M. | A.M. | P.M. | A.M. | P.M. |
| 1 | 2.0 | 2.0 | 2.0 | 2.1 | 2.1 | 2.0 | 1.9 | 1.9 | 2.0 | 2.0 |
| 2 | 2.0 | 2.1 | 2.1 | 1.9 | 1.9 | 1.9 | 1.9 | 2.0 | 2.0 | 2.0 |
| 3 | 2.1 | 2.1 | 2.1 | 3.2 | 2.1 | 2.0 | 2.1 | 2.0 | 2.1 | 2.0 |
| 4 | 1.9 | 1.9 | 2.0 | 3.7 | 2.0 | 2.0 | 2.0 | 2.0 | 2.0 | 2.0 |
| 5 | 2.0 | 2.0 | 2.1 | 3.6 | 2.0 | 2.1 | 1.9 | 1.0 | 1.9 | 2.1 |
| **Mean:** | **2.00** | **2.02** | **2.06** | **2.90** | **2.02** | **2.00** | **1.96** | **1.78** | **2.00** | **2.02** |
| **Range:** | **0.2** | **0.2** | **0.1** | **1.8** | **0.2** | **0.2** | **0.2** | **1.0** | **0.2** | **0.1** |

In looking at the averages of the subgroups, it can be seen that they are close to the requirement of 2.0 in. with the exception of four subgroups: the Tuesday A.M. subgroup, which averaged 2.06 in.; the Tuesday P.M. subgroup, which averaged 2.9 in.; the Thursday A.M. subgroup of 1.96 in.; and the Thursday P.M. subgroup of 1.78 in. In addition, when looking at the ranges of the subgroups, the Tuesday P.M. subgroup is very inconsistent with a range of 1.8 in., and the Thursday P.M. subgroup demonstrates inconsistency as well with a range of 1.0 in.

The grand average, which is the centerline for the X-bar chart, is now computed by taking the average of all of the subgroup averages:

$$\overline{\overline{X}} = \frac{(2.0+2.02+2.06+2.9+2.02+2.0+1.96+1.78+2.0+2.02)}{10} = 2.076 \text{ in.}$$

Likewise, the grand range, which is the centerline for the R-chart, is computed by taking the average of all of the subgroup ranges:

$$\overline{R} = \frac{(0.2+0.2+0.1+0.8+0.2+0.2+0.2+1.0+0.2+0.1)}{10} = 0.42 \text{ in.}$$

With the calculations now computed, the upper and lower control limits can be calculated. To accomplish this, the control chart factors are used to compute the upper and lower control limits for both the average and range charts. Table 3 in the appendix lists the factors for different subgroup sizes. Computing the upper and lower control limits for each of the charts is shown below:

AVERAGE CHART:  $\text{UCL}_X = \overline{\overline{X}} + A_2 \overline{R} = 2.076 + (0.577)(0.42) = 2.32 \text{ in.}$
$\text{LCL}_X = \overline{\overline{X}} - A_2 \overline{R} = 2.076 - (0.577)(0.42) = 1.84 \text{ in.}$

RANGE CHART:  $\text{UCL}_R = D_4 \overline{R} = (2.114)(0.42) = 0.89 \text{ in.}$
$\text{LCL}_R = D_3 \overline{R} = (0.000)(0.42) = 0 \text{ in.}$

The last step is to construct the actual control chart, as shown in Figure 9–2. Note that both the X-bar and R-charts have data points that lie above the upper control line. This situation (out of control) is described and discussed in Section 9.6 of this chapter.

## 9.5.2 Average and Sigma Charts

The average and sigma charts are very similar to the X-bar and R-charts with two philosophical differences:

1. The range is replaced by the sample standard deviation of the subgroup. The sample standard deviation is called sigma and is represented by the lower case s.
2. The X-bar and s-charts are used when subgroups of usually ten or more items exist.

Although the sigma chart is more accurate than the range chart, it does have short-falls in that it is more difficult to compute, understand, and interpret. Notwithstanding, s is the sample standard deviation and when subgroup sizes are larger than 10, s is a much better estimate of the standard deviation than R. This is due to the fact that when the data are not close in proximity and extreme values occur, s is much less affected by the extreme values than R will be. The following formula is used when s is computed:

$$s = \sqrt{\frac{\Sigma x^2 - (\Sigma x)^2 / n}{n-1}}$$

As with the X-bar and R-charts, control chart factors exist to make the construction of the chart simpler.

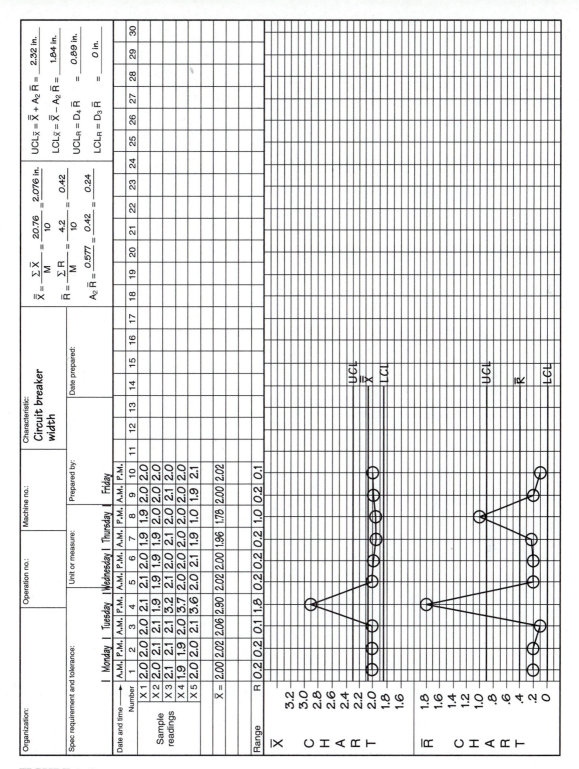

**FIGURE 9–2**

X-Bar and R-Chart for Circuit Breaker Problem

**EXAMPLE 1**   A continuous three-shift, seven-days-per-week automated process mixes various chemicals together for resale to vendors. Data on the amount of a certain chemical in the mixture will be monitored by taking ten samples every hour for twelve continuous hours. The following data in Table 9–4 represent the parts per liter of the chemical that has been recorded. In addition, the mean and s for each subgroup have been calculated and are shown below the raw data.

**TABLE 9–4**
Data for Chemical Mix Problem

| Sample No. | 7 A.M. | 8 A.M. | 9 A.M. | 10 A.M. | 11 A.M. | 12 noon | 1 P.M. | 2 P.M. | 3 P.M. | 4 P.M. | 5 P.M. | 6 P.M. |
|---|---|---|---|---|---|---|---|---|---|---|---|---|
| 1 | 1.13 | 1.14 | 1.13 | 1.15 | 1.14 | 1.20 | 1.57 | 1.21 | 1.17 | 1.18 | 1.20 | 1.11 |
| 2 | 1.14 | 1.12 | 1.13 | 1.20 | 1.09 | 1.06 | 1.22 | 1.11 | 1.10 | 1.15 | 1.21 | 1.15 |
| 3 | 1.10 | 1.22 | 1.14 | 1.16 | 1.12 | 1.10 | 1.09 | 1.09 | 1.17 | 1.14 | 1.16 | 1.14 |
| 4 | 1.11 | 1.14 | 1.17 | 1.11 | 1.14 | 1.19 | 1.15 | 1.14 | 1.13 | 1.12 | 1.12 | 1.10 |
| 5 | 1.17 | 1.14 | 1.12 | 1.18 | 1.20 | 1.18 | 1.16 | 1.12 | 1.22 | 1.20 | 1.11 | 1.17 |
| 6 | 1.15 | 1.15 | 1.17 | 1.11 | 1.08 | 1.11 | 1.14 | 1.17 | 1.19 | 1.11 | 1.16 | 1.14 |
| 7 | 1.16 | 1.13 | 1.14 | 1.07 | 1.17 | 1.07 | 1.13 | 1.15 | 1.17 | 1.12 | 1.11 | 1.18 |
| 8 | 1.10 | 1.10 | 1.12 | 1.11 | 1.15 | 1.14 | 1.14 | 1.16 | 1.10 | 1.19 | 1.10 | 1.20 |
| 9 | 1.07 | 1.15 | 1.10 | 1.08 | 1.06 | 1.16 | 1.13 | 1.12 | 1.11 | 1.17 | 1.12 | 1.11 |
| 10 | 1.06 | 1.10 | 1.11 | 1.14 | 1.12 | 1.10 | 1.10 | 1.10 | 1.06 | 1.13 | 1.14 | 1.16 |
| **Mean** | **1.12** | **1.14** | **1.13** | **1.13** | **1.13** | **1.13** | **1.18** | **1.14** | **1.14** | **1.15** | **1.14** | **1.15** |
| s | .037 | .034 | .023 | .042 | .042 | .050 | .140 | .040 | .050 | .032 | .039 | .033 |

As with the X-bar and R-chart, the grand mean and grand s (which are the centerlines for the charts) are now calculated in the same manner as the grand mean and grand range were in the X-bar and R-chart.

$$\overline{\overline{X}} = \frac{(1.12 + 1.14 + 1.13 + 1.13 + 1.13 + 1.13 + 1.18 + 1.14 + 1.14 + 1.15 + 1.14 + 1.15)}{12}$$

$$\overline{\overline{X}} = 1.14 \text{ parts per liter}$$

$$\overline{s} = \frac{(.037 + .034 + .023 + .042 + .042 + .050 + .140 + .040 + .050 + .032 + .039 + .033)}{12}$$

$$\overline{s} = .047 \text{ part per liter}$$

And finally, before the construction of the actual chart, the upper and lower control limits need to be computed for the Average and s-charts. There are control chart factors that exist for the Average and s-chart just like the X-bar and R-chart. And just like the X-bar and R-chart, these factors are based on the subgroup size.

AVERAGE CHART:

$$UCL_X = \overline{\overline{X}} + A_3\,\overline{s} = 1.14 + (0.975)(.047) = 1.19 \text{ parts per liter}$$
$$LCL_X = \overline{\overline{X}} - A_3\,\overline{s} = 1.14 - (0.975)(.047) = 1.09 \text{ parts per liter}$$

s CHART:

$$UCL_s = B_4\,\overline{s} = (1.716)(.047) = 0.081 \text{ parts per liter}$$
$$LCL_s = B_3\,\overline{s} = (0.284)(.047) = 0.013 \text{ parts per liter}$$

Now that all of the calculations have been performed, the control chart is drawn, as shown in Figure 9–3.

## 9.5.3 Median and Range Charts

Median and range charts are similar to X-bar and R-charts and the X-bar and s-charts. One of the reasons for their development is the fact that they are mathematically much less complicated than their counterparts. In fact, they are easy to understand and relatively quick to construct. It is logical to realize that people in industry and government may shy away from the concept of control charts because of the mathematics involved. Thus, it may be beneficial for a company or organization to start with the median and range charts as a way to introduce the concept of control charting. Even so, although the median and range charts are easier to construct, the downside to these charts is that from a mathematical point of view they do not contain nearly as much information as the X-bar and R-charts and the X-bar and s-charts. With this in mind, the median and range charts may be used as an introduction to statistical process control, but progression to the other more powerful and useful charts should occur after individuals become comfortable with the concepts.

EXAMPLE 1    A new type of fertilizer spreader has been developed and is currently being tested before final mass production. Tests are being conducted because the nozzle of the dispenser is made of a new material that may or may not be as reliable as the type of nozzle used on older conventional types of spreaders. The test will measure the amount of fertilizer released over a five-minute time span at five different times each day for a two-week period. The data, measured in pounds of fertilizer, were recorded and are shown in Table 9–5. Likewise, the median (the middle value when the data are arranged in ascending order) and range have been computed for each subgroup.

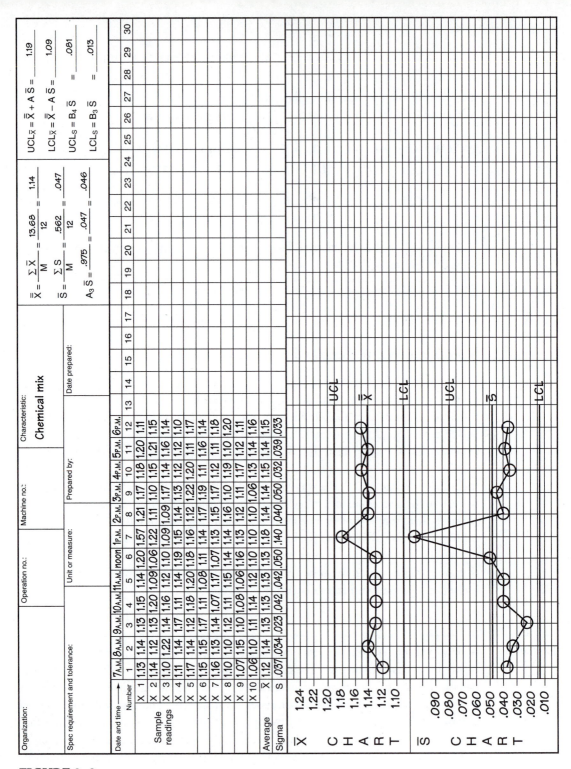

**FIGURE 9–3**

X-Bar and s-Chart for Chemical Mix Problem

**TABLE 9–5**

Data for Fertilizer Spreader Problem

| Sample No. | M | T | W | TH | F | M | T | W | TH | F |
|---|---|---|---|---|---|---|---|---|---|---|
| 1 (8 A.M.) | 8.5 | 8.5 | 8.6 | 8.7 | 8.4 | 8.2 | 8.5 | 8.4 | 8.7 | 8.3 |
| 2 (10 A.M.) | 8.3 | 8.6 | 8.3 | 8.4 | 8.5 | 8.2 | 8.7 | 8.8 | 8.4 | 8.7 |
| 3 (1 P.M.) | 8.9 | 8.4 | 8.8 | 8.5 | 8.3 | 8.9 | 8.4 | 8.8 | 8.2 | 8.3 |
| 4 (3 P.M.) | 8.8 | 8.5 | 8.6 | 8.4 | 8.9 | 8.5 | 8.4 | 8.7 | 8.1 | 8.9 |
| 5 (5 P.M.) | 8.1 | 8.3 | 8.5 | 8.4 | 8.8 | 8.6 | 8.1 | 8.4 | 8.5 | 8.1 |
| **Median** | **8.5** | **8.5** | **8.6** | **8.4** | **8.5** | **8.5** | **8.4** | **8.7** | **8.4** | **8.3** |
| **Range** | **0.8** | **0.3** | **0.5** | **0.3** | **0.6** | **0.7** | **0.6** | **0.5** | **0.6** | **0.8** |

In continuing the standard procedure for building control charts, the next step is the computation of the centerlines for the median chart and the range chart:

$$\overline{\overline{Me}} = \frac{8.5+8.5+8.6+8.4+8.5+8.5+8.4+8.7+8.4+8.3}{10}$$

$$\overline{\overline{Me}} = 8.48 \text{ pounds}$$

$$\overline{R} = \frac{0.8+0.3+0.5+0.3+0.6+0.7+0.6+0.5+0.6+0.8}{10}$$

$$\overline{R} = 0.57 \text{ pounds}$$

Before drawing the control charts, the upper and lower control limits for the median and range charts need to be computed. The computation of the control limits for the range chart is the same as it was for the range chart in the X-bar and R-chart:

MEDIAN CHART: $\text{UCLMe} = \overline{\overline{Me}} + A_6 \overline{R} = 8.48 + (0.691)(0.57) = 8.87$ pounds

$\text{LCLMe} = \overline{\overline{Me}} + A_6 \overline{R} = 8.48 - (0.691)(0.57) = 8.09$ pounds

RANGE CHART: $\text{UCL}_R = D_4 \overline{R} = (2.114)(0.57) = 1.20$ pounds
$\text{LCL}_R = D_3 \overline{R} = (0)(0.57) = 0$ pounds

The control chart is now ready to be drawn, as shown in Figure 9–4.

It should be apparent by now that the procedure for building a variables control chart is not dependent on which control chart is selected for usage. The only things that change in the procedure are the control chart factors for computing upper and lower control limits, and the factors are dependent on the type of chart that is used. The rest of the procedure is the same, and this is true as well for the last variables control chart (individuals and range chart) to be discussed in this chapter.

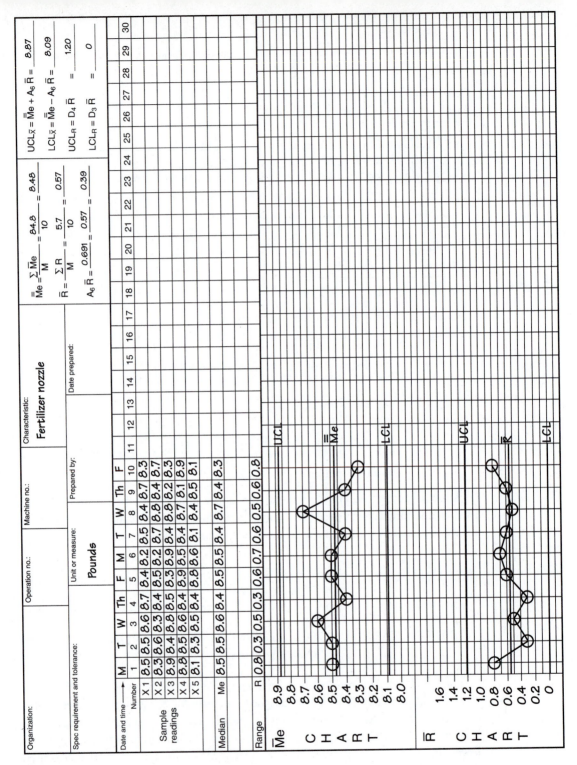

**FIGURE 9–4**
Median and Range Chart for Fertilizer Spreader Problem

## 9.5.4 Individuals and Range Charts

There may be times when the use of subgroups is not beneficial. A good example is one that involves destructive testing. For example, to destructively test ten subgroups of five products may not be economical or logical. Likewise, there may also be times when long periods of time occur between each data point. For example, data may only be available for collection once every quarter of a year, which means that there will only be four data points per year. Just because instances like these occur does not mean that the concept of statistical process control needs to be abandoned. Rather, it is instances like these that merely dictate that the data be looked at on an individual basis as opposed to a subgroup. It is these types of situations where the individuals chart is extremely helpful.

With an individuals chart, the ability to compute a range is impossible because there is no subgroup, but only individual data. Notwithstanding, the moving range is used to compensate for this. The moving range is computed by taking the absolute value of the current observation minus the previous observation. The control chart factor for the upper and lower control limits of the individuals chart varies slightly as well, which will be seen in the following example. Other than these exceptions, the steps for the individuals chart is the same as that for the other variables charts.

As a final note, in several of the examples in this chapter on variables control charts ten subgroups were used. There is no magical number about the ten subgroups. The ten subgroups are for simplicity to demonstrate the technique.

---

**EXAMPLE 1**   The management of a local solid waste dumping facility is interested in the amount of trash that is being received from Route 8. This route is located in the suburbs and while the square mileage of the route is constant, the number of customers is increasing due to several new condominium and housing initiatives. Trash pickup on this route occurs once a week and it is decided to collect data for ¼ of a year to study the situation. The data, in tons of trash per week, are recorded and shown in Table 9–6, along with the moving range (recall that the moving range is the absolute value of the current value minus the previous value).

**TABLE 9–6**
Data for Trash Collection Problem

| Week No. | Amount of Trash | Moving Range | Week No. | Amount of Trash | Moving Range |
|---|---|---|---|---|---|
| 1 | 6.3 | — | 14 | 6.6 | 0.1 |
| 2 | 6.1 | 0.2 | 15 | 6.7 | 0.1 |
| 3 | 6.4 | 0.3 | 16 | 6.5 | 0.2 |
| 4 | 6.5 | 0.1 | 17 | 6.3 | 0.2 |

*(continued)*

**TABLE 9–6** *(cont'd.)*
Data For Trash Collection Problem

| Week No. | Amount of Trash | Moving Range | Week No. | Amount of Trash | Moving Range |
|---|---|---|---|---|---|
| 5 | 6.0 | 0.5 | 18 | 6.9 | 0.6 |
| 6 | 6.8 | 0.8 | 19 | 6.7 | 0.2 |
| 7 | 6.6 | 0.2 | 20 | 6.6 | 0.1 |
| 8 | 6.8 | 0.2 | 21 | 6.2 | 0.4 |
| 9 | 6.4 | 0.4 | 22 | 6.7 | 0.5 |
| 10 | 6.4 | 0.0 | 23 | 6.9 | 0.2 |
| 11 | 6.7 | 0.3 | 24 | 6.7 | 0.2 |
| 12 | 6.8 | 0.1 | 25 | 7.0 | 0.3 |
| 13 | 6.5 | 0.3 | 26 | 6.8 | 0.2 |
| | | | TOTAL: | 170.9 | 6.7 |

The centerlines for both the individuals and range chart are now calculated:

$$\overline{X} = \frac{170.9}{26}$$

$$\overline{X} = 6.57 \text{ tons per week}$$

$$\overline{R} = \frac{6.7}{25}$$

$$\overline{R} = 0.268 \text{ tons per week}$$

As with the other variables control charts, the upper and lower control limits can now be computed for the individuals and range chart. The control limits for the range chart are calculated the same way as they are in the X-bar and R-chart and the median and range chart:

INDIVIDUALS CHART:  $UCL_x = \overline{x} + E_2 \overline{R} = 6.57 + (2.66)(0.268) = 7.28$ tons per week
$LCL_x = \overline{x} - E_2 \overline{R} = 6.57 - (2.66)(0.268) = 5.86$ tons per week

NOTE: The value for E2 in this example was for a subgroup of size 2. This is because two values were used to compute each of the moving ranges.

RANGE CHART:       $UCL_R = D_4 \overline{R} = (3.267)(0.268) = 0.88$ ton per week
$LCL_R = D_3 \overline{R} = (0.000)(0.268) = 0.00$ ton per week

The Individuals and Range Charts are shown in Figure 9–5.

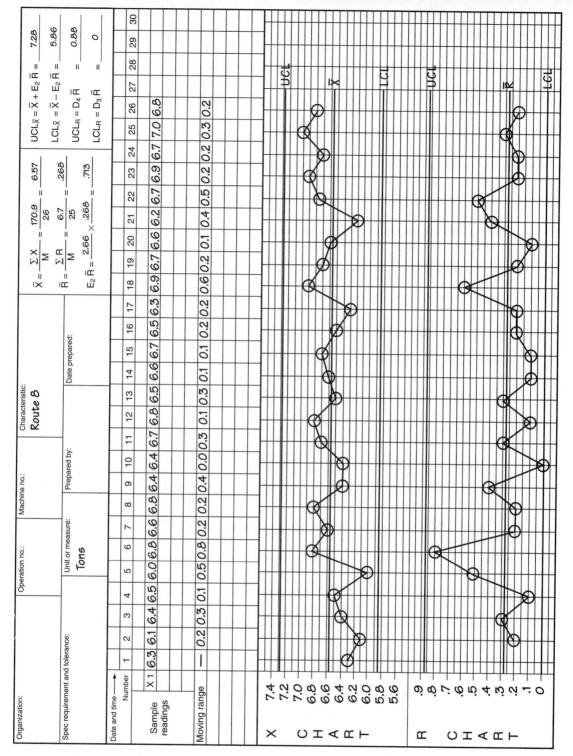

**FIGURE 9–5**

Individual and Range Chart for Trash Collection Problem

## 9.6 Control Chart Interpretation

Up to now, the types of variables control charts and their construction have been the center of discussion. Now it is time to look at the interpretation of control charts. The shapes (or patterns) of the control chart give insight into possible investigative areas for the maintaining of statistical control and potential process improvement. It is important to remember that it is not always necessary to take action because any of the patterns occur in the control chart, especially if the process is meeting or exceeding the customer specification. Rather, the control chart patterns are used to generate proactive investigation that may result in process improvement. Tampering with a process is over control that will cause negative results. The patterns identified in this section are not unique to variables control charts. They are valid for attributes control charts as well, and are potential signals to investigate the process. For explanation purposes of the various patterns described next, it is assumed that the Y-axis indicates the number scrapped of a process and thus higher numbers are worse than lower numbers. The graphical representation of the various control chart patterns explained next is shown in Figure 9–6.

- Natural Pattern (Statistical Control). This occurs when there are no points lying outside the control limits. There are no runs, cycles, or other patterns. Although some of the points may approach the control limits, most of the points (approximately 67% or ⅔) of the points lie within the mean ± 1 standard deviation. All variation is common cause and no action is required.
- Out of Control. One point lies above the upper control limit or below the lower control limit. Investigate the point for special cause variation. Adjustment to the process may or may not be necessary.
- Consecutive Points Near the Control Limit. When two or more consecutive points lie very close to the control limit, this may be a sign to investigate the cause. Two situations of this pattern can occur:

  1. Two Consecutive Points Near Upper Control Limit. Possible cause for the investigation of poor performance.
  2. Two Consecutive Points Near Lower Control Limit. Possible cause for the investigation of improved performance.

- Run. A run results when consecutive points lie above or below the centerline. Runs are usually investigated when they reach seven or eight consecutive points. Two types of runs can occur:

  1. Run above Centerline. Indicative of sustained poor performance. Investigate causes of poor performance, especially if the process is performing above the customer specification.
  2. Run below Centerline. Indicative of sustained process improvement.

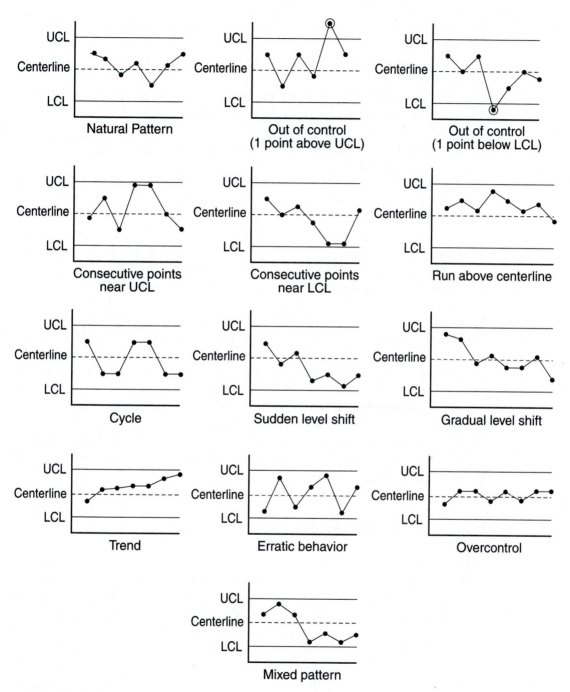

**FIGURE 9–6**
Control Chart Patterns

- Cycles. Cycles are regularly repeating patterns of ups and downs exhibited by a process. Shift start-ups and changes in personnel are examples of situations that would cause cyclical behavior. Adjustment to the process may or may not be required.
- Level Shifts. The level of a pattern can shift. It may be a sudden or a gradual shift, but the level changes and is noticeable on the control chart:

  1. Sudden Level Shift. This is easily detectable on the control chart. The process average has shifted.
  2. Gradual Level Shift. This is usually indicative of some portion of the process being changed. A good example is changing the procedure of a process and a learning curve is being experienced.

- Trends. A trend is seven or more consecutive points that are either increasing or decreasing. A trend is usually indicative of some process change.
- Unusual or Erratic Behavior. This pattern occurs when very few points lie near the centerline and are widely spaced all over the control chart but within the control limits.
- Overcontrol. This pattern occurs when approximately seven consecutive points are very close to the centerline. When this occurs, the supervision of the process needs to be notified to investigate and take corrective action. Overcontrol typically refers to taking corrective action when none is needed, or there is no indication of an out of control situation. When overcontrol occurs, it will usually produce approximately 150% of the normal variation of the process.
- Mixed Patterns. This occurs when two or more distributions in the process are present. An example would be material being used in process and the material is being supplied by two different suppliers.

## 9.7 Document and Rebaseline the Process

As with every step in the roadmap, documentation and rebaselining must occur. Recall that this step includes quantitative data regarding the central tendency and variability of the process. The nice thing about variables control charts is that once an improvement has been made to the process the chart has the inherent rebaseline statistics embodied in it. The chart is made using the grand average (central tendency) and grand range (variability). Once a new improvement initiative has been implemented and results in improvement over time, a new centerline and control limits are established. The change between the old centerlines and control limits and the new centerlines and control limits is the quantification of the process improvement.

**EXAMPLE 1**   A process improvement initiative has been under way on a particular product line. The initiative included control charting using the Average and Range charts on the number of items that were required to be reworked. The control charts were put into place and data collected over time to control chart the process. Once the process control charts had enough data to profile the process, the group interpreted the control charts and was able to identify an improvement to the process. The improvement was implemented and control charting continued. After some time, the control chart centerlines and control limits were reestablished to quantify the actual improvement in terms of central tendency and variability to the process. The before and after control charts are shown in Figure 9–7.

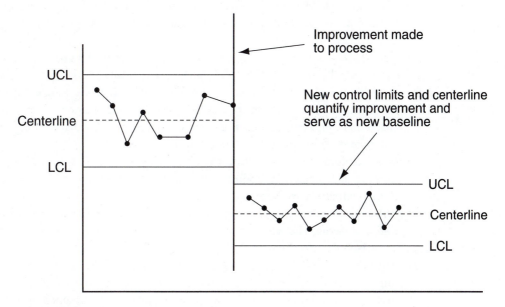

**FIGURE 9–7**
Documentation and Rebaselining with Control Charts

## 9.8 Summary

- A variables control chart is used when the data of the process are of the measuring type.
- The four basic types of variables control charts are: Average and Range chart, Average and Sigma chart, Median and Range chart, and Individuals and Range chart.

- Subgroup size is a key prerequisite to control charts and the size of the subgroup should be sufficient to provide for not only variation within the group, but variation between the groups as well.
- Control chart factors are factors that are used in the computation of the upper and lower control limits for the different types of variables control charts.
- The Average and Range chart is the most powerful and popular of the variables control charts.
- The Average and Sigma chart is similar to the Average and Range chart with the exception that the sigma denotes the sample standard deviation of the subgroup. Likewise, the Average and Sigma control chart is usually used when subgroups consist of ten or more items.
- The Sigma chart is more powerful than the range when large subgroup sizes exist because when extreme values occur, the standard deviation will be much less affected than the range.
- The Median and Range chart is an excellent chart to start an organization with due to the fact that it is relatively simple (mathematically speaking) to construct than the other variables control charts. Although the Median and Range chart is good to introduce personnel to control charting concepts, progression to the other control charts is recommended. This is due to the fact that the median is not as good of an indicator of central tendency as the average.
- When circumstances such as destructive testing to obtain data, or large intervals of time occur between data points are available to be gathered, the Individuals and Range chart is appropriate to use.
- Although there are different types of variables control charts to use, it is the situation that depicts the type of control chart to use. Variables data are still the focus of all variables control charts.
- After the control chart has been constructed, interpretation is necessary to determine if various patterns exist. Patterns serve to point out areas in the process that may need investigation due to poor performance or sustained superior performance. The main types of patterns that exist are:

  - Normal pattern
  - Out of control
  - Consecutive points near the control limit
  - Runs
  - Cycles
  - Level shifts
  - Trends
  - Unusual or erratic behavior
  - Overcontrol
  - Mixed patterns

# Review Questions

1. The number of basic variables control charts is:
   a. one
   b. two
   c. three
   d. four
   e. none of the above

2. The most popular and most used type of variable control chart is the:
   a. Average and Range chart
   b. Average and Sigma chart
   c. Median and Range chart
   d. Individuals and Range chart
   e. none of the above

3. A control chart that is recommended for usage when subgroups of ten or more exist is the:
   a. Average and Range chart
   b. Average and Sigma chart
   c. Median and Range chart
   d. Individuals and Range chart
   e. none of the above

4. The control chart that is used when long intervals of time exist between data collection points is the:
   a. Average and Range chart
   b. Average and Sigma chart
   c. Median and Range chart
   d. Individuals and Range chart
   e. none of the above

5. The type of control chart that is a good chart to start an organization with because it is easy to compute (mathematically speaking) is the:
   a. Average and Range chart
   b. Average and Sigma chart
   c. Median and Range chart
   d. Individuals and Range chart
   e. none of the above

6. When one or more data points lie above the upper control limit or below the lower control limit, this is indicative of which type of control chart pattern?
   a. Trend
   b. Run

    c. Cycle
    d. Out of control
    e. None of the above

7. Regularly repeating up and down patterns is indicative of which type of control chart pattern?
    a. Mixed pattern
    b. Overcontrol
    c. Cycle
    d. Out of control
    e. None of the above

8. A control chart factor is a factor which:
    a. is dependent on the subgroup size
    b. helps determine the centerline of the control chart
    c. is sometimes used to assist in the calculation of the upper and lower control limits
    d. a and c
    e. none of the above

9. Seven or more consecutive points above or below the centerline is indicative of which type of control chart pattern?
    a. In control
    b. Run
    c. Level shift
    d. Out of control
    e. None of the above

10. This type of pattern occurs when approximately seven or more consecutive points are very close to the centerline.
    a. Trend
    b. Run
    c. Overcontrol
    d. In control
    e. None of the above

11. This type of pattern exists when very few points lie near the centerline, and are widely spaced all over the control chart but within the control limits.
    a. Unusual or erratic behavior
    b. In control
    b. Out of control
    c. Overcontrol
    d. None of the above

12. Consecutive increasing or decreasing points are indicative of which type of control chart pattern?
    a. Level shift
    b. Cycle
    c. Run
    d. In control
    e. None of the above

13. Some reasons for using variables control charts are:
    a. to identify when the process should be left alone and no action taken
    b. to monitor the process
    c. to satisfy management
    d. all of the above
    e. a and b
    f. none of the above

14. Two or more types of patterns that exist in the process is a description of which type of control chart pattern?
    a. Cycle
    b. Trend
    c. Out of control
    d. Mixed pattern
    e. None of the above

15. The type of control chart pattern that describes the situation where there are no points that lie outside of the control limits is:
    a. run
    b. out of control
    c. level shift
    d. cycle
    e. none of the above

16. The Average and Sigma charts use which control chart factors for the computation of the upper and lower control limits?
    a. A3, B3, and B4
    b. A6, D3, and D4
    c. E2, D3, and D4
    d. A2, D3 and D4
    e. None of the above

17. The Average and Range charts use which control chart factors for the computation of the upper and lower control limits?
    a. A3, B3, and B4
    b. A6, D3, and D4

c. E2, D3, and D4
d. A2, D3 and D4
e. None of the above

18. The Individuals and Range charts use which control charts for the computation of the upper and lower control limits?
a. A3, B3, and B4
b. A6, D3, and D4
c. E2, D3, and D4
d. A2, D3 and D4
e. None of the above

19. The Median and Range charts use which control chart factors for the computation of the upper and lower control limits?
a. A3, B3, and B4
b. A6, D3, and D4
c. E2, D3, and D4
d. A2, D3 and D4
e. None of the above

20. Control chart centerlines are computed using:
a. standard deviation
b. grand average
c. both a and b
d. none of the above

# Problems

1. A certain dimension on a part from a manufacturing process is being studied. Twenty samples of five parts yield the results in Table 9–7. Construct an Average and Range chart for the data in the table.

**TABLE 9–7**
Manufacturing Process Data

| Sample | Average | Range |
|--------|---------|-------|
| 1 | 8 | 9 |
| 2 | 8 | 5 |
| 3 | 13 | 8 |
| 4 | 7 | 4 |
| 5 | 11 | 7 |

**TABLE 9–7** *(cont'd.)*

| Sample | Average | Range |
|--------|---------|-------|
| 6 | 9 | 6 |
| 7 | 6 | 9 |
| 8 | 10 | 6 |
| 9 | 10 | 8 |
| 10 | 7 | 4 |
| 11 | 5 | 5 |
| 12 | 6 | 3 |
| 13 | 11 | 6 |
| 14 | 12 | 9 |
| 15 | 9 | 5 |
| 16 | 12 | 7 |
| 17 | 14 | 6 |
| 18 | 8 | 6 |
| 19 | 10 | 8 |
| 20 | 7 | 5 |

2. A slurry mixture for making magnets consists of a portion of water. It is important to ensure that the amount of water be relatively consistent because if there is not enough water in the mixture, then the magnets will crack when dried in the kiln. Likewise, if there is too much water in the slurry mixture, then the magnets will not form correctly. Data are gathered (in gallons of water per slurry mixture) and recorded as shown in Table 9–8. For these data, construct an Average and Sigma chart.

**TABLE 9–8**
Magnet Slurry Mixture Data

| Sample No. | M | T | W | TH | F |
|------------|-----|-----|-----|-----|-----|
| 1 | 6.0 | 6.1 | 6.2 | 6.3 | 6.4 |
| 2 | 5.9 | 6.1 | 6.3 | 5.9 | 6.1 |
| 3 | 6.0 | 6.0 | 6.0 | 6.4 | 6.0 |
| 4 | 6.2 | 6.3 | 6.4 | 6.5 | 6.7 |
| 5 | 5.5 | 5.7 | 6.4 | 6.0 | 6.0 |
| 6 | 5.2 | 6.0 | 6.0 | 6.0 | 6.1 |
| 7 | 5.9 | 5.8 | 5.7 | 5.9 | 5.9 |
| 8 | 6.0 | 6.1 | 5.9 | 5.9 | 6.0 |
| 9 | 6.1 | 6.2 | 6.0 | 6.0 | 6.0 |
| 10 | 6.3 | 6.3 | 6.5 | 6.6 | 5.7 |

3. A process is in statistical control and the upper and lower tolerance limits are within the control limits of the process. Management is confused with the fact that scrap occurs when the process is in statistical control. How would you explain the situation to them? What will your recommendation be to them on how to reduce the amount of scrap produced by the process?

4. A destructive test is conducted on fifteen products with the hopes of gaining insight into the performance of the process. The data collected are shown in Table 9–9. For the data shown, construct an Individuals and Range chart.

**TABLE 9–9**
Destructive Test Data

| Sample No. | Data |
|------------|------|
| 1 | 35 |
| 2 | 32 |
| 3 | 33 |
| 4 | 37 |
| 5 | 36 |
| 6 | 35 |
| 7 | 35 |
| 8 | 30 |
| 9 | 43 |
| 10 | 39 |
| 11 | 31 |
| 12 | 31 |
| 13 | 29 |
| 14 | 38 |
| 15 | 37 |

5. It is decided that control charts are going to be introduced in a new manufacturing facility at critical control points throughout the process. The workforce has exhibited strong resistance to the concept of control charts because they feel that control charts are too complicated and take too much time, thus preventing them from their primary job of manufacturing. Also, because the new plant was employing an incentive pay system for workers based on their individual production quotas of quality products, the workers felt that taking time out to chart a process could be better spent producing the items. Management understood the concerns of the workers, but got the workforce to agree to participate in a trial control charting initiative on one product line. If they were comfortable with the concept and understood how it would help them meet their goals of quality

products, they would agree to begin participating in a plantwide process control program. Data were collected from the test product line and are shown in Table 9–10. As a member of management, you have been selected to demonstrate the simplicity of control charting. You decide on the Median and Range chart as a beginning point. With your decision being made, construct the Median and Range chart for the data in Table 9–10.

**TABLE 9–10**
Manufacturing Plant Trial Data

| Sample No. | M | T | W | TH | F |
|---|---|---|---|---|---|
| 1 | 72 | 73 | 73 | 76 | 81 |
| 2 | 75 | 76 | 71 | 70 | 75 |
| 3 | 79 | 76 | 72 | 75 | 77 |
| 4 | 76 | 75 | 72 | 74 | 78 |
| 5 | 74 | 74 | 71 | 75 | 79 |
| 6 | 75 | 72 | 72 | 73 | 80 |
| 7 | 76 | 77 | 73 | 76 | 78 |
| 8 | 74 | 75 | 73 | 77 | 80 |

# 10

## Step 4B: Attributes Control Charts

### 10.1 Chapter Objectives

After completing this chapter, the student should be able to:

- Understand the concepts of attributes control charts.
- Know what attributes control charts are and which attributes control chart to use under various situations.
- Know the two categories of control charts and which control charts belong in each category.
- Know how to construct the four different types on attributes control charts.
- Understand the relationship between variables and attributes control charts.
- Revisit control chart interpretation and learn some additional patterns that may serve as potential areas of investigation. These new patterns are a result of subdividing the control chart into regions based on the standard deviation.

### 10.2 Key Terms

| | | |
|---|---|---|
| Area of Opportunity Control Charts | c Charts | np Charts |
| Binomial Count Control Charts | Defective Unit | p Charts |
| | Nonconforming Unit | u Charts |
| | Nonconformity | |

## **10.3** Introduction

In Chapter 9 variables control charts were explained as one part of Step 4. The other part of Step 4 is attributes control charts. There may be times during implementation of Step 4 of the roadmap that only variables control charts will fit the particular situation and are used. Likewise, there may be times when only attributes control charts fit the particular situation and are used. And there are still other times when both variables and attributes control charts may be used. Recall that control charts are of great assistance in process improvement, and that it is the situation that determines which

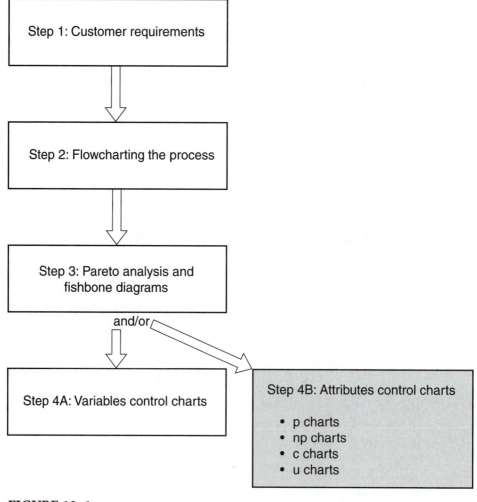

**FIGURE 10–1**
Roadmap Step 4B

control chart should be used to gain insight into the process. As with variables control charts, attributes control charts are very powerful in pointing to the area(s) of the process that may need to be investigated for opportunities for improvement.

The main difference between the two types of control charts is the type of data that are being collected. The attributes control charts deal with data that have to do with the number of times a particular event occurs, whereas variables control charts deal with data that involve measurement.

Figure 10–1 depicts the roadmap to date and highlights the attributes control chart portion of Step 4.

## 10.4  Attributes Control Chart Overview

As with variables control charts and their key concepts, there are some key concepts associated with attributes control charts that must be understood before commencing. Correct understanding of these concepts allows for correct selection of which attributes chart to use, as well as correct interpretation of the chart itself.

### 10.4.1 Attributes Nomenclature

It is definitely necessary to review some nomenclature before starting. The best way to do this is through an example.

---

EXAMPLE 1    A company produces snow blowers. Each snow blower is considered a unit of production. Within that unit of production, a **nonconformity** may exist. A nonconformity is some departure, or deviation, of a quality characteristic from its specification. There may be only one nonconformity in the unit of production, or there may be several nonconformities. A **nonconforming unit** is defined as a product containing at least one nonconformity.

When evaluating in terms of the ability to use a unit of production, the word **defective unit** comes into play. A unit is considered defective when the unit contains one or more nonconformities. However, it should be noted that there are situations where a unit of production is considered not defective while possessing one or more nonconformities. For example, suppose that the snow blower has a very slight run in the paint that is barely visible to the eye. This type of nonconformity would probably be classified as a minor nonconformity, and in terms of usability the unit would probably be classified as not defective. Suppose however that the snow blower has a severe crack over the casing around the chain. This type of nonconformity would probably be classified as a major nonconformity, and the unit of production would probably be considered defective.

## 10.4.2 Advantages of Attributes Control Charts

Recall from Chapter 9 that variables control charts tell a great deal of information about the product or process because they involve very specific measurement of the product or process. As a result of this specific measurement, they are very sensitive to what is happening in the process. Therefore, variables control charts have the reputation of being more sensitive than attributes charts, and this is demonstrated in Section 10.4.3.

Notwithstanding, attributes control charts have some distinct advantages as well. One advantage is that they provide a mechanism for quick analysis and summary of the process or product quality. For example, there may be times when management is interested in the amount of scrap or rework associated with a particular product. Likewise, a plant manager may be interested in the categorization of a product into the classifications of conformance to engineering specifications and nonconformance to engineering specifications. These types of situations are excellent examples of where the attributes control charts prove very beneficial.

In addition to the examples stated above, there is also a second advantage of the attributes control chart. Recall that the Median and Range control chart was recommended as an introduction step for the workforce to variables control charting, with advancement to the more sophisticated variables control charts after proficiency on Median and Range charts was obtained. Attributes control charts are generally much easier to understand and explain than variables control charts, especially when the audience consists of personnel who either are not familiar with control charts or do not use them on a regular basis. This ease of understanding is a distinct advantage associated with attributes control charts.

A third advantage of attributes control charts is that data collection for these charts may not be as complicated as with variables charts. For example, it may be quicker to classify products into the two categories of conforming to engineering specifications and nonconforming to engineering specifications as opposed to in-depth measurement that may be involved with constructing variables control chart on the products.

A final advantage is that attributes control charts may be the only type of chart that is appropriate for some situations. Some specific examples may include the percent errors in accounts receivable, percent errors in accounts payable, percent on-time shipments, and percent on-time departures.

## 10.4.3 Classifications of Attributes Control Charts

There are two different classifications of attributes charts that are discussed in this chapter: binomial count control charts and area of opportunity control charts.

> **Binomial count control charts**: this type of attributes control chart deals with either the number or fraction of items in subgroups that meet a particular event or have a particular characteristic.

**Area of opportunity control charts**: these attributes control charts are based on the Poisson distribution and deal with the number of times a particular event or particular characteristic occurs in a given area of opportunity.

From the classification of attributes control charts, the sensitivity of attributes control charts versus variables control charts can be further seen and demonstrated. For example, in the usage of binomial charts, where the data fall into one of two groups (usually pass or fail), variables control charts will reveal more information about the process. This is because classification into the pass or fail categories is much quicker and less rigorous than the measurement that would be associated for variables control chart purposes.

## 10.4.4 Types of Attributes Charts

There are several types of attributes charts and similar to variables control charts, each chart has its own purpose. The four basic types of attributes control charts that are addressed in this text are:

- **p charts**
- **np charts**
- **c charts**
- **u charts**

## 10.4.5 Steps in Making an Attributes Control Chart

The construction steps are basically the same for all of the attributes control charts. It is only the calculations for the grand average and control limits that will vary depending on which control chart is used. The basic steps in attributes control chart construction are shown in Table 10–1.

**TABLE 10–1**
Steps in Constructing an Attributes Control Chart

| |
|---|
| STEP 1: Select the process to be studied. |
| STEP 2: Determine which attributes control chart will best fit the process. |
| STEP 3: Determine a subgroup size that will be large enough to obtain a representative count. |
| STEP 4: Collect the data on a data collection sheet. |
| STEP 5: Calculate the centerline for the chart. |
| STEP 6: Compute the control limits for the chart. |
| STEP 7: Graph the chart. |

# 10.5 Step 4B: Attributes Control Charts

The specific attributes control charts (n, np, c, and u) are now discussed and applications demonstrated. As with variables control charts, it is important to know the difference between the charts and for which situation each chart will apply to.

## 10.5.1 p Charts

The p chart is a binomial control chart used to measure proportions, the proportion nonconforming, or the percent of items that meet a particular event. Specifically, a p chart is used to determine the percent of nonconforming items in a sample. When using a p chart, the subgroup size may be constant or may vary:

1. p chart for constant subgroup sizes: constant subgroup sizes are used when a discrete countable characteristic exists (for example, number of failures on a weekly quiz given to a class of students).
2. p chart for variable subgroup sizes: variable subgroup sizes are a little more difficult than the straightforward p chart where subgroup sizes are constant. However, there are times when this type of situation cannot be prevented. There are three ways to solve the varying subgroup size issue. They include averaging the subgroup size, varying control limits, and including two sets of control limits. Although all three methodologies are viable, this text uses the average subgroup size as the preferred methodology for handling variable subgroup sizes. As a side note, if the individual subgroups are larger than 125% of the average or less than 75% of the average, then the control limits should be recalculated for those subgroups only.

The centerline (average percent nonconforming) and upper and lower control limits for the p chart are computed via the following formulas:

$$\bar{p} = \text{Total number nonconforming} \,/\, \text{total number inspected}$$

$$\text{Upper Control Limit} = \bar{p} + 3\left[\bar{p}\left(1-\bar{p}\right)/n\right]^{\frac{1}{2}}$$

$$\text{Lower Control Limit} = \bar{p} - 3\left[\bar{p}\left(1-\bar{p}\right)/n\right]^{\frac{1}{2}}$$

where:   $n$ = sample size (for constant subgroup sizes
$n$ = average sample size (for varying subgroup sizes)

---

**EXAMPLE 1**   (p chart for constant subgroup sizes): A manufacturing plant produces lightbulbs. The bulbs arrive at the end of the line and are then automatically packaged into boxes that hold twenty bulbs. It is the policy of the plant to set every sixth box aside for inspection and nonconforming unit identification. Data are collected on a worksheet and shown in Table 10–2.

**TABLE 10–2**
Lightbulb Nonconforming Data

| Sample No. | Sample Size | No. Nonconforming | % Nonconforming |
|---|---|---|---|
| 1 | 20 | 2 | 10.0 |
| 2 | 20 | 0 | 0.00 |
| 3 | 20 | 4 | 20.00 |
| 4 | 20 | 0 | 0.00 |
| 5 | 20 | 0 | 0.00 |
| 6 | 20 | 0 | 0.00 |
| 7 | 20 | 1 | 5.00 |
| 8 | 20 | 2 | 10.00 |
| 9 | 20 | 3 | 15.00 |
| 10 | 20 | 0 | 0.00 |
| 11 | 20 | 0 | 0.00 |
| 12 | 20 | 1 | 5.00 |
| 13 | 20 | 0 | 0.00 |
| 14 | 20 | 0 | 0.00 |
| 15 | 20 | 0 | 0.00 |
| 16 | 20 | 2 | 10.00 |
| 17 | 20 | 0 | 0.00 |
| 18 | 20 | 1 | 5.00 |
| 19 | 20 | 1 | 5.00 |
| 20 | 20 | 0 | 0.00 |

The next step is to compute the average percent nonconforming.

$$\bar{p} = \frac{2+0+4+0+0+0+1+2+3+0+0+1+0+0+0+2+0+1+1+0}{400}$$

$$\bar{p} = 0.0425$$
$$\bar{p} = 4.25\%$$

The next and final step before constructing the control chart is to compute the upper and lower control limits.

$$UCL = \bar{p} + 3\left[\bar{p}(1-\bar{p})/n\right]^{\frac{1}{2}} \qquad \text{where } n = 20$$

$$UCL = 0.0425 + 3\left[0.0425(1-0.0425)/20\right]^{\frac{1}{2}}$$

$$UCL = 0.1778$$

$$LCL = \bar{p} - 3\left[\bar{p}(1-\bar{p})/n\right]^{\frac{1}{2}}$$

$$LCL = 0.0425 - 3\left[0.0425(1-0.0425)/20\right]^{\frac{1}{2}}$$

$$LCL = -0.0928$$

Before constructing the chart there is an important item to point out. When dealing with percents and proportions, the mathematical calculation associated with the control limits may result in a UCL above 1.00 (or 100%) and an LCL below 0.00 (or 0.00%). When thinking about this, it is illogical to ever have more than 100% nonconforming units or less than 0% nonconforming units. Thus, in attributes control charts instances like this, it is permissible and recommended to round the UCL down to 1.00 (or 100%) and round the LCL up to 0 (or 0%). In this example, the LCL is rounded to 0.

The p chart, shown in Figure 10–2, can now be drawn.

In looking at the control chart and comparing it to the various control chart patterns, a pattern can be observed. The process went out of statistical control on the third box of lightbulbs. As stated in Chapter 9, this pattern serves as a potential area of investigation. Another item to consider in continuing attributes control charting is that because more than 50% of the subgroups show 0% nonconforming, a larger subgroup size should be used. This is because the subgroup size should be large enough that there is a good chance of finding at least one nonconforming in any subgroup.

---

**EXAMPLE 2**    (p chart with unequal subgroup sizes): A new portion of a manufacturing line has been installed to produce oil filters that are used on automobiles. This portion of the line was purchased because the tooling was aging and becoming very inaccurate and inconsistent with respect to the inside threads that mount the filter to the engine. Because this portion of the line is brand new, it is desired by management to get a rough idea as to the quality of the inside threads on the oil filter in terms of meeting or not meeting engineering specifications. It is decided that a p chart will give an idea as to the initial quality of the new portion of the line. Furthermore, a sample will be obtained every hour by going to the line and emptying the bin that finished filters are placed in. These filters will then be taken to the quality lab and will serve as the basis

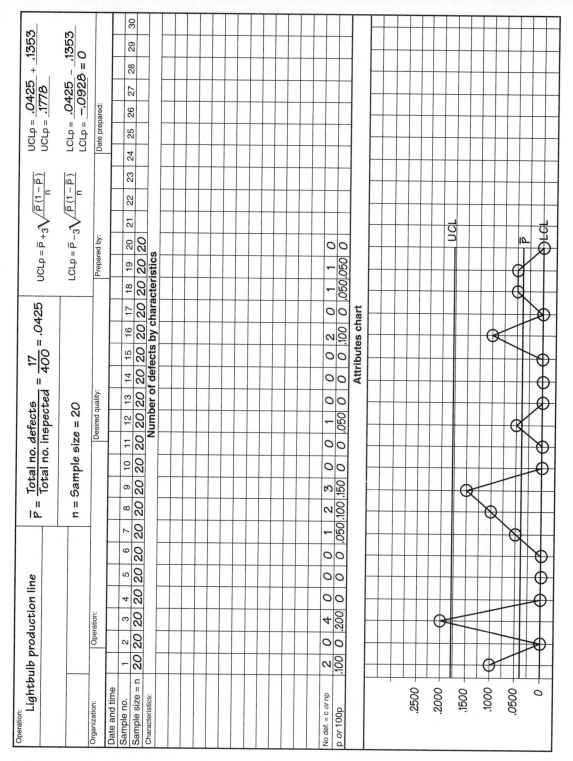

**FIGURE 10–2**

p Chart for Lightbulb Problem

for the p chart. The bin does not have the same number of oil filters in it each hour, thus the sample size will not be the same. Samples are taken every hour for a three-shift period and the data collection results are shown in Table 10–3.

## TABLE 10–3
Oil Filter Nonconforming Data

| Sample No. | Sample Size | No. Nonconforming | % Nonconforming |
|---|---|---|---|
| 1 | 75 | 2 | 2.67 |
| 2 | 80 | 4 | 5.00 |
| 3 | 80 | 2 | 2.50 |
| 4 | 85 | 0 | 0.00 |
| 5 | 78 | 0 | 0.00 |
| 6 | 79 | 1 | 1.27 |
| 7 | 80 | 2 | 2.50 |
| 8 | 84 | 0 | 0.00 |
| 9 | 81 | 0 | 0.00 |
| 10 | 82 | 0 | 0.00 |
| 11 | 83 | 1 | 1.20 |
| 12 | 85 | 5 | 5.88 |
| 13 | 75 | 6 | 8.00 |
| 14 | 75 | 5 | 6.67 |
| 15 | 77 | 1 | 1.30 |
| 16 | 84 | 1 | 1.19 |
| 17 | 79 | 0 | 0.00 |
| 18 | 80 | 0 | 0.00 |
| 19 | 81 | 1 | 1.23 |
| 20 | 84 | 0 | 0.00 |
| 21 | 79 | 0 | 0.00 |
| 22 | 78 | 0 | 0.00 |
| 23 | 78 | 1 | 1.28 |
| 24 | 80 | 0 | 0.00 |

Now that the data have been collected, the average percent nonconforming can be computed.

$$\bar{p} = \frac{2+4+2+0+0+1+2+0+0+0+1+5+6+5+1+1+0+0+1+0+0+0+1+0}{1922}$$

$$\bar{p} = 0.0167$$
$$\bar{p} = 1.67\%$$

The next step is to compute the upper and lower control limits. Note that the value of the subgroup size (n) will be the average subgroup size.

$$UCL = \overline{p} + 3\left[\overline{p}(1-\overline{p})/n\right]^{\frac{1}{2}} \qquad \text{where } n = 1922/24 = 80.08$$

$$UCL = 0.0167 + 3\left[0.0167(1-0.0167)/80.08\right]^{\frac{1}{2}}$$

$$UCL = 0.0597$$

$$LCL = \overline{p} - 3\left[\overline{p}(1-\overline{p})/n\right]^{\frac{1}{2}}$$

$$LCL = 0.0167 - 3\left[0.0167(1-0.167)/80.08\right]^{\frac{1}{2}}$$

$$LCL = -0.0368 = 0.00 \quad \text{(Remember from Example 1 that in the}$$
application of attributes control charts you cannot have less than 0.00% nonconforming. Thus, the LCL is rounded to 0.00.)

Now that the calculations have been completed, the p chart can be drawn, as shown in Figure 10–3.

When looking at the control chart with respect to patterns, it can be seen that there are a couple of areas for investigation:

- The thirteenth and fourteenth points are out of control
- The ten-point run below the LCL (fifteenth through twenty-fourth point)

## 10.5.2 np Charts

The np chart is another binomial chart. It is used to count the number of nonconforming items in a uniform or nonvarying sample size. In other words, the count is an integer that represents the number of nonconforming items in a constant sample, or subgroup. Thus, the np chart uses counts while the p chart uses percents, or fractions. The same steps are used to construct the np chart that are used to construct the p chart, with the exception that the centerline is np as opposed to the average percent nonconforming. The centerline is calculated as follows:

np = (subgroup size)(total number nonconforming/total number inspected)

The control limits are computed as follows:

$$UCL = np + 3\left[np(1-p)\right]^{\frac{1}{2}}$$

$$LCL = np - 3\left[np(1-p)\right]^{\frac{1}{2}}$$

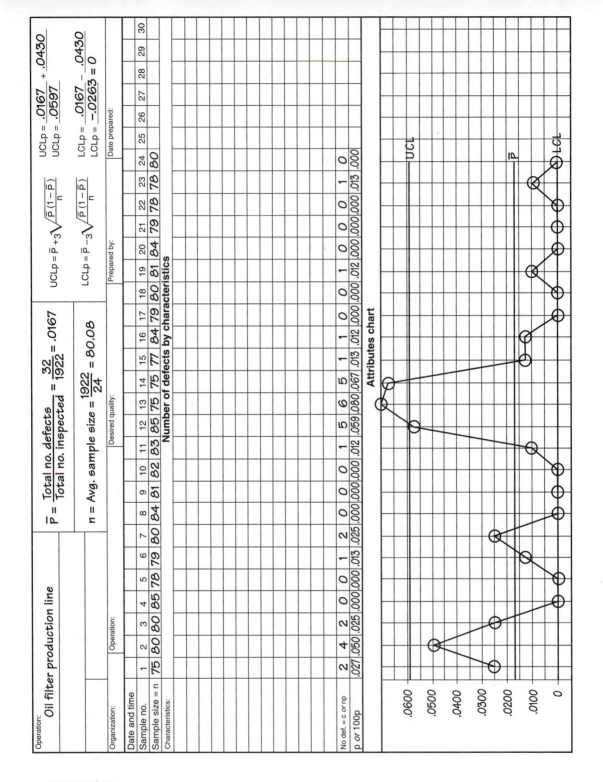

**FIGURE 10–3**

p Chart for Oil Filter Problem

272

**EXAMPLE 1**  Bricks are manufactured for use in housing construction. The bricks are produced and then put on pallets for shipment. Samples of fifty are taken each hour and visually inspected for cracks, breaking, and other flaws that would result in construction problems. The data for one three-shift operation day is shown in Table 10–4.

**TABLE 10–4**
Brick Nonconforming Data

| Sample No. | Sample Size | No. Nonconforming | % Nonconformance |
|:---:|:---:|:---:|:---:|
| 1 | 50 | 0 | 0.00 |
| 2 | 50 | 0 | 0.00 |
| 3 | 50 | 1 | 0.02 |
| 4 | 50 | 1 | 0.02 |
| 5 | 50 | 3 | 0.06 |
| 6 | 50 | 0 | 0.00 |
| 7 | 50 | 2 | 0.04 |
| 8 | 50 | 1 | 0.02 |
| 9 | 50 | 4 | 0.08 |
| 10 | 50 | 1 | 0.02 |
| 11 | 50 | 1 | 0.02 |
| 12 | 50 | 2 | 0.04 |
| 13 | 50 | 3 | 0.06 |
| 14 | 50 | 0 | 0.00 |
| 15 | 50 | 0 | 0.00 |
| 16 | 50 | 5 | 0.10 |
| 17 | 50 | 1 | 0.02 |
| 18 | 50 | 1 | 0.02 |
| 19 | 50 | 3 | 0.06 |
| 20 | 50 | 0 | 0.00 |
| 21 | 50 | 2 | 0.04 |
| 22 | 50 | 0 | 0.00 |
| 23 | 50 | 1 | 0.02 |
| 24 | 50 | 3 | 0.06 |

Now that the nonconforming data have been collected, the calculation steps for construction of the np chart can be performed, starting with the centerline (np):

np = (subgroup size)(total number of nonconforming/total number inspected)

$$np = (50)(35/1200) = 1.458$$

The upper and lower control limits are now computed as follows:

$$UCL = np + 3\left[np(1-p)\right]^{\frac{1}{2}}$$

$$UCL = 1.458 + 3\left[1.458(1-0.0292)\right]^{\frac{1}{2}}$$

$$UCL = 5.027$$

$$LCL = np - 3\left[np(1-p)\right]^{\frac{1}{2}}$$

$$LCL = 1.458 - 3\left[1.458(1-0.0292)\right]^{\frac{1}{2}}$$

$$LCL = -2.111 = 0$$

As with the p charts, the LCL is rounded to 0 because it is impossible to have less than 0 nonconforming.

With the computations for the np chart completed, the actual control chart can now be drawn, as shown in Figure 10–4.

When looking at the control chart with respect to the control chart patterns, it can be seen that the process is relatively normal.

## 10.5.3 c Charts

Area of opportunity charts deal with the number of times a particular event or particular characteristic occur in a given area of opportunity. An example is that of the automobile, where one or more defects can occur in the given area of opportunity (the automobile). There are two types of area of opportunity charts: the c chart and the u chart.

The c chart is an area of opportunity control chart that counts the number of non-conformities per unit. Areas of opportunity that are constant in size means that a constant subgroup size is present. The c chart is best used when there is a small number of errors in a sample space and the nonconforming events are independent. Ten cubic yards of topsoil, one particular model of automobiles, and 5 square yards of wallpaper are examples of when the c chart is used and the subgroup size is constant. c is the number of events (or nonconformities) per each area of opportunity. The average of c's over time is the centerline for the c chart and is computed as follows:

$$\bar{c} = \text{Total number of events} / \text{Number of areas of opportunity}$$

As with all of the other attributes control charts, although the steps for construction are the same, the individual calculations for control limits vary as well.

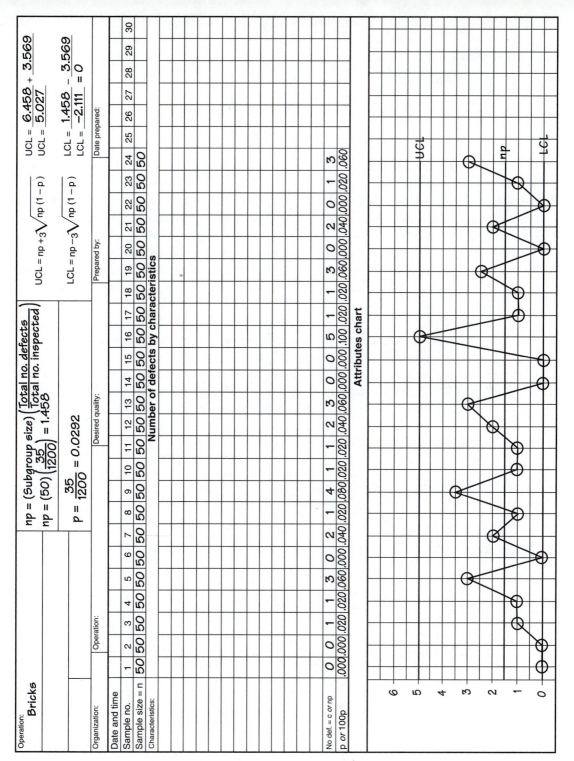

**FIGURE 10–4**
np Chart for Brick Problem

$$UCL = \bar{c} + 3\left[\bar{c}\right]^{\frac{1}{2}}$$

$$LCL = \bar{c} - 3\left[\bar{c}\right]^{\frac{1}{2}}$$

**EXAMPLE 1**    Flat 2 ft. wide × ¼ in. thickness steel is produced and rolled on a spool, or roll. Each spool consists of 25 ft. of steel and after the spools are banded, they are shipped to various manufacturers. The spool is usually then straightened out, put on a punch press, and then continuously fed into the punch press where various stampings are punched. At the production plant for the steel, a worker sits at the end of the line and inspects the number of nonconformities (in this case, deep pits) per spool of steel. Experience has shown that customers have returned spools of steel after discovering too many deep pits when the steel is unrolled, straightened out, and fed to the punch press. The entire process of producing a spool of steel and shipping it to the customer only to have it shipped back as unacceptable is extremely costly. Thus, it is important to ensure that quality spools of steel are produced, and c charts are used for each steel production line to ensure the desired high quality. A typical data collection sheet for a three-shift day for one of the production lines is shown in Table 10–5.

## TABLE 10–5
Steel Nonconformity Data

| Spool No. | No. Deep Pits |
|-----------|---------------|
| 1M | 5 |
| 2M | 8 |
| 3M | 3 |
| 4M | 9 |
| 5M | 6 |
| 6M | 3 |
| 7M | 8 |
| 8M | 4 |
| 9M | 4 |
| 10M | 10 |
| 11M | 5 |
| 12M | 4 |
| 13M | 4 |
| 14M | 7 |
| 15M | 2 |
| 16M | 3 |
| 17M | 2 |

**TABLE 10–5** *(cont'd.)*

| Spool No. | No. Deep Pits |
|-----------|---------------|
| 18M | 5 |
| 19M | 6 |
| 20M | 4 |
| 21M | 9 |

From the data collection sheet, the centerline and upper and lower control limits for the c chart can be calculated as follows:

$$\bar{c} = \text{Total Number of Deep Pits} / \text{Number of Areas of Opportunity}$$

$$\bar{c} = 111 / 21 = 5.286$$

$$\text{UCL} = \bar{c} + 3[\bar{c}]^{\frac{1}{2}}$$

$$\text{UCL} = 5.286 + 3[5.286]^{\frac{1}{2}}$$

$$\text{UCL} = 12.183$$

$$\text{LCL} = \bar{c} - 3[\bar{c}]^{\frac{1}{2}}$$

$$\text{LCL} = 5.286 - 3[5.286]^{\frac{1}{2}}$$

$$\text{LCL} = -1.611 = 0 \quad \text{And again, because the negative number is}$$
meaningless for the LCL, it is rounded to 0.

The c chart for the steel production line is shown in Figure 10–5.

When looking for patterns in the steel spool control chart, it can be seen that the process appears to exhibit no special control chart patterns.

## 10.5.4 u Charts

When the area of opportunity varies, then the u chart is used. The u chart is very similar to the c chart with the exception of varying subgroup sizes. A good example of varying areas of opportunity would be the number of typographical errors in a document. Each document size, or area of opportunity, varies. Another example is the number of nonconformities that occur in sections of material cut from a huge roll of material used in the clothing industry. Sections are manually cut with a scissors and are not all the same size. As with the document size, each section, or area of opportunity, will vary.

Similar to the c chart, the u chart counts events (for example, nonconformities) that occur in a given area of opportunity. But with the u chart, the events are expressed as a

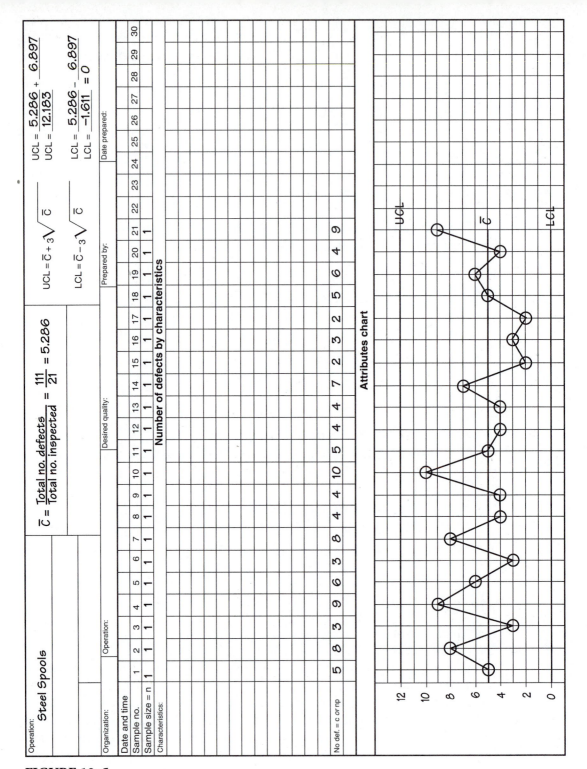

**FIGURE 10–5**
c Chart for Steel Problem

278

percent of the quantity (or size) of the area of opportunity. For instance, in the clothing example where sections are cut from the roll of material, the c chart considers the number of defects per roll of material. The u chart, on the other hand, considers the number of defects per some size (for example, defects per square foot of material). Each section, even though they vary, has a number of defects associated with it.

The centerline and upper and lower control limits for the u chart are computed differently than they are for the c chart. The centerline for the u chart is the total number of nonconformities of all subgroups divided by the sum of all of the individual areas of opportunities for all subgroups:

$$\bar{u} = \frac{\text{Total number of nonconformities of all subgroups}}{\text{Total of areas of opportunity of all subgroups}}$$

And in mathematical symbols:

$$\bar{u} = \Sigma c_i / \Sigma a_i$$

Because the subgroup size varies, the upper and lower control limits will vary from point to point. Such as there were three different possible solutions to accommodate varying control limits with the p chart, there are two different solutions to choose from to accommodate for the varying subgroup sizes associated with the u chart. One solution is to compute the control limits based on averaging the values for the area of opportunity. The other solution is to adopt variable control limits. Variable control limits are more confusing, difficult, and time consuming to compute and apply. As with the three p chart solutions where average subgroup size solution was adopted, the technique that is applied and demonstrated in this text is that of averaging the values for the area of opportunity. With this technique in mind, the computation for the upper and lower control limits becomes:

$$\text{UCL} = \bar{u} + 3\left[\bar{u} / \bar{a}\right]^{\frac{1}{2}}$$

$$\text{LCL} = \bar{u} - 3\left[\bar{u} / \bar{a}\right]^{\frac{1}{2}}$$

---

**EXAMPLE 1**   Rolls of various gage electrical wire are supplied in a tool crib shop in a plant. The wire is cut on a daily basis by the crib attendant to the various lengths required by job shop workers in the plant. After the attendant has cut the wire he inspects it for severe nicks and cuts. A piece of wire that has severe nicks and cuts will not be used in the final product. A u chart is kept for each gage wire and the nonconformity data for a roll of 14 gage wire is shown in Table 10–6.

**TABLE 10–6**
Electrical Wire Nonconformity Data

| Sample No. | Linear Feet of Wire | Area of Opportunity (5 linear ft.), a | Number of Nonconformities c | Nonconformities Per 5 ft. |
|---|---|---|---|---|
| 1 | 15 | 3 | 6 | 2.00 |
| 2 | 20 | 4 | 3 | 0.75 |
| 3 | 5 | 1 | 0 | 0.00 |
| 4 | 10 | 2 | 1 | 0.50 |
| 5 | 10 | 2 | 3 | 1.50 |
| 6 | 25 | 5 | 5 | 1.00 |
| 7 | 25 | 5 | 4 | 0.80 |
| 8 | 15 | 3 | 1 | 0.33 |
| 9 | 10 | 2 | 5 | 2.50 |
| 10 | 20 | 4 | 0 | 0.00 |
| 11 | 5 | 1 | 0 | 0.00 |
| 12 | 30 | 6 | 6 | 1.00 |
| 13 | 25 | 5 | 3 | 0.60 |
| 14 | 40 | 8 | 12 | 1.50 |
| 15 | 10 | 2 | 1 | 0.50 |
| 16 | 10 | 2 | 0 | 0.00 |
| 17 | 25 | 5 | 2 | 0.40 |
| 18 | 35 | 7 | 8 | 1.14 |
| 19 | 30 | 6 | 5 | 0.83 |
| 20 | 10 | 2 | 1 | 0.50 |
| 21 | 15 | 3 | 2 | 0.67 |
| 22 | 20 | 4 | 3 | 0.75 |

From the raw data, the centerline and control limits can now be computed:

$$\bar{u} = \Sigma c_i / \Sigma a_i = 71 / 82 = 0.866$$

$$UCL = \bar{u} + 3[\bar{u} / \bar{a}]^{\frac{1}{2}}$$

$$UCL = 0.866 + 3[0.866 / 3.727]^{\frac{1}{2}}$$

where $\bar{a}$ = the average area of opportunity = $\Sigma a_i / n = 82 / 22 = 3.727$

$$UCL = 2.312$$

$$LCL = \bar{u} - 3[\bar{u} / \bar{a}]^{\frac{1}{2}}$$

$$LCL = 0.866 - 3[0.866 / 3.727]^{\frac{1}{2}}$$

$LCL = -0.580 = 0$ And again, because it is impossible to have less than 0 nonconformities, the LCL is rounded to 0.

The u chart, shown in Figure 10–6, can now be drawn.

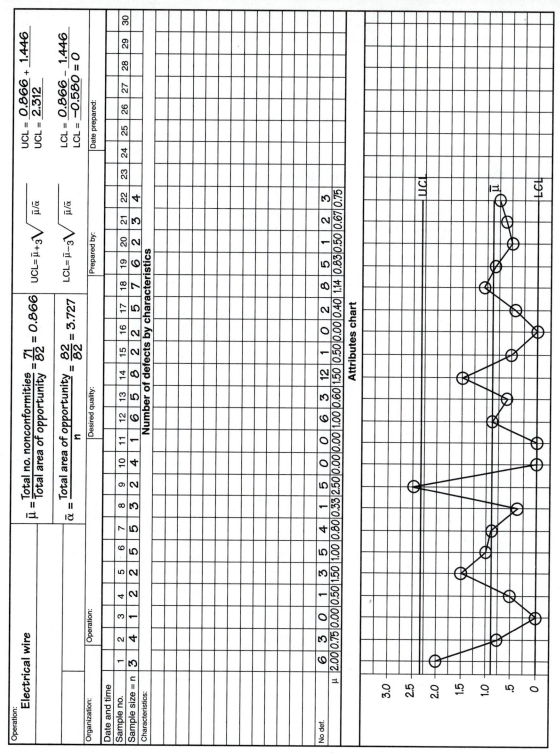

**FIGURE 10–6**
u Chart for Electrical Wire Problem

281

The u chart for the electrical wire has several areas that can be investigated. First and foremost, the process is out of control and the ninth data point needs to be analyzed for the root cause.

## 10.6 Control Chart Interpretation

As shown in the preceding examples, the same control chart patterns for variables charts (reference Figure 9–6) are also used for attributes control charts. But it is again important to remember that these patterns are general guidelines and are intended to point to potential areas for investigation. The patterns are not absolutes.

In addition, as the application of statistical process control becomes resident in a plant and is an ongoing implemented philosophy, there are further advanced guidelines that can be applied to control charts. These guidelines usually involve further subdividing the control chart into regions based on the standard deviation, as shown in Figure 10–7.

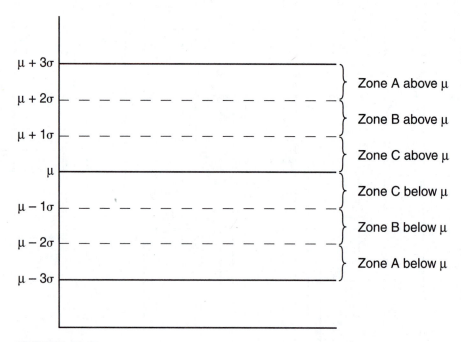

**FIGURE 10–7**
Zones for Patterns

With the control chart now broken into zones, more specific guidelines may be applied for possible areas of investigation and action that may or may not be required. Examples of the patterns and what they mean include:

- Two out of three points in a row in Zone A or beyond serves as a possible warning of a process shift.
- Fifteen points in a row in Zone C means that there is smaller variability in the process than would normally be expected.
- Six points in a row that are steadily increasing or decreasing is usually indicative of the process drifting. This is usually caused by such things as tool wear or aging machinery.
- Seven or nine points in a row steadily increasing or decreasing usually signals that the process average has changed.
- Fourteen points in a row that alternate up and down is usually associated with two different causes present in the process. An example would be two suppliers of materials to one manufacturing line, and the two suppliers of materials are not exactly the same.

## 10.7 Summary

- Attributes control charts deal with data that have to do with the number of times a particular event occurs.
- Attributes control charts are classified in two main categories: binomial count control charts and area of opportunity control charts.
- Binomial count control charts deal with either the number or fraction of items in a subgroup that meet a particular event or have a particular characteristic.
- Area of opportunity control charts deal with the number of times a particular event or characteristic occurs in a given area of opportunity.
- p Charts are binomial count control charts that are used for proportions, or to measure the proportion nonconforming.
- p Charts can be used when the subgroup is uniform or variable.
- np Charts are binomial count control charts that count the number of nonconforming items in a uniform or nonvarying sample size or subgroup.
- c Charts are area of opportunity control charts that count the number of nonconformities per unit.
- u Charts are area of opportunity control charts that are similar to the c chart except that the subgroup size varies.
- The same control chart patterns that are used for variables control charts are also used for attributes control charts. These control chart patterns serve as an indicator

for possible areas for investigation. It must be remembered, however, that just because a pattern exists, action may or may not be required.

- After statistical process control charts are resident and routinely used, other advanced methods for pattern recognition within control charts include dividing the control chart area into zones based on the standard deviation.

## Review Questions

1. Attributes control charts can be classified in the following into:
   a. binomial and areas of opportunity
   b. binomial and only constant areas of opportunity
   c. binomial and only varying areas of opportunity
   d. none of the above

2. The binomial attributes control charts discussed in this chapter are:
   a. p chart and c chart
   b. p chart and u chart
   c. u chart and c chart
   d. np chart and u chart
   e. none of the above

3. The areas of opportunity control charts discussed in this chapter are:
   a. np chart and p chart
   b. u chart and c chart
   c. np chart and u chart
   d. c chart and np chart
   e. none of the above

4. Attributes charts are excellent for:
   a. detailed information about individual product measurements
   b. classification of the product into pass and fail categories
   c. obtaining a relatively quick idea as to the quality of a product group
   d. a and b
   e. b and c
   f. none of the above

5. The binomial control chart that is used when the sample size is constant is the:
   a. p chart
   b. u chart
   c. np chart
   d. c chart
   e. none of the above

6. The area of opportunity chart that can handle different subgroup sizes is the:
   a. u chart
   b. np chart
   c. c chart
   d. p chart
   e. none of the above

7. If you were after the percent nonconforming in a varying subgroup size, the attributes control chart that best fits this situation is the:
   a. u chart
   b. np chart
   c. p chart
   d. c chart
   e. none of the above

8. If you were after the number of nonconforming items in a series of standard lot sizes, the attributes control chart that best fits the situation is the:
   a. u chart
   b. np chart
   c. p chart
   d. c chart
   e. none of the above

9. A control chart that deals with the number of times a given event occurs in a given area is:
   a. area of opportunity control chart
   b. p chart
   c. np chart
   d. none of the above

10. A control chart that would be used to measure the number of nonconformities in incoming rolls of steel that are delivered to the plant is the:
    a. binomial control chart
    b. n chart
    c. c chart
    d. np chart
    e. none of the above

11. Attributes control charts typically require more measurement than variables control charts. True or False

12. Binomial control charts typically require more measurement than areas of opportunity control charts. True or False

13. The control chart that would be used to count the number of nonconformities per stereo is the:
    a. np chart
    b. c chart
    c. u chart
    d. p chart
    e. a variables control chart

14. When a p chart is used and the subgroup size varies, the method that can be used is:
    a. average subgroup size
    b. three sets of control limits
    c. nonvarying control limits
    d. none of the above

15. A unit of production can contain nonconformities and the unit still be classified as usable. True or False

16. Attributes control charts are generally more sensitive than variables control charts. True or False

17. It is highly recommended that the attributes control charts be used before variables control charts are used in a particular situation. True or False

18. The main difference between binomial control charts and area of opportunity charts is that:
    a. the binomial control charts cannot handle different subgroup sizes
    b. the area of opportunity control charts cannot handle different subgroup sizes
    c. there is no difference
    d. none of the above

19. The same types of patterns (for example, runs, cycles, and level shifts) that apply to variables control charts for interpretation also apply to attributes control charts. True or False

20. Control chart patterns that divide the region into zones based on the standard deviation are only used on attributes control charts. True or False

## Problems

1. A company has designed a production line for a new model of computer speakers. The design of the speakers is such that if they do not operate correctly after being built they are scrapped as opposed to reworked due to the extremely high cost of diagnosis and rework. Management is interested in getting an idea of the scrap

associated with the process. The quality department decides to take a sample of fifteen sets of speakers per hour for a 24-hour period to determine the percent of scrap produced by the line. The data are shown in Table 10–7.

**TABLE 10–7**

Computer Speaker Nonconforming Data

| Sample No. | Sample Size | No. Nonconforming | % Nonconformance |
|---|---|---|---|
| 1 | 15 | 3 | 20.00 |
| 2 | 15 | 0 | 0.00 |
| 3 | 15 | 2 | 13.33 |
| 4 | 15 | 2 | 13.33 |
| 5 | 15 | 0 | 0.00 |
| 6 | 15 | 0 | 0.00 |
| 7 | 15 | 2 | 13.33 |
| 8 | 15 | 1 | 6.67 |
| 9 | 15 | 1 | 6.67 |
| 10 | 15 | 0 | 0.00 |
| 11 | 15 | 0 | 0.00 |
| 12 | 15 | 0 | 0.00 |
| 13 | 15 | 1 | 6.67 |
| 14 | 15 | 1 | 6.67 |
| 15 | 15 | 0 | 0.00 |
| 16 | 15 | 0 | 0.00 |
| 17 | 15 | 3 | 20.00 |
| 18 | 15 | 3 | 20.00 |
| 19 | 15 | 0 | 0.00 |
| 20 | 15 | 0 | 0.00 |
| 21 | 15 | 0 | 0.00 |
| 22 | 15 | 3 | 20.00 |
| 23 | 15 | 1 | 6.67 |
| 24 | 15 | 1 | 6.67 |

a. Construct the p chart.
b. Referring to the patterns used for variables and attributes control charts both in Chapter 9 and this chapter, discuss any patterns that may be resident in the process.

2. For the p chart in problem 1, what are the upper and lower boundaries for:
   a. Zone A above the centerline
   b. Zone B above the centerline

    c. Zone C above the centerline

    d. Zone A below the centerline

    e. Zone B below the centerline

    f. Zone C below the centerline

3. A process needs to be studied for process improvement. Because the process has never been studied or has never been the subject of any control chart monitoring, the process improvement group decides as an initial step to learn more about the process by gathering some data, constructing a p chart, and analyzing the control chart for any patterns. Data are collected on the process and are shown in Table 10–8.

**TABLE 10–8**

Process Nonconforming Data

| Sample No. | Sample Size | No. Nonconforming | % Nonconformance |
|---|---|---|---|
| 1 | 10 | 0 | 0.00 |
| 2 | 12 | 1 | 8.33 |
| 3 | 10 | 1 | 10.00 |
| 4 | 11 | 0 | 0.00 |
| 5 | 9 | 2 | 22.22 |
| 6 | 13 | 0 | 0.00 |
| 7 | 14 | 3 | 21.43 |
| 8 | 12 | 3 | 25.00 |
| 9 | 9 | 2 | 22.22 |
| 10 | 10 | 2 | 20.00 |
| 11 | 12 | 0 | 0.00 |
| 12 | 13 | 0 | 0.00 |
| 13 | 15 | 0 | 0.00 |
| 14 | 9 | 1 | 11.11 |
| 15 | 14 | 0 | 0.00 |
| 16 | 8 | 0 | 0.00 |
| 17 | 8 | 0 | 0.00 |
| 18 | 9 | 0 | 0.00 |
| 19 | 11 | 1 | 9.09 |
| 20 | 10 | 0 | 0.00 |

    a. Construct a p chart for the data in Table 10–8.

    b. Interpret the control chart for any patterns discussed in this chapter or Chapter 9.

4. For the p chart in problem 3, what are the boundaries for:
   a. Zone A above the centerline
   b. Zone B above the centerline
   c. Zone C above the centerline
   d. Zone A below the centerline
   e. Zone B below the centerline
   f. Zone C below the centerline

5. Steel rods are produced and plated at a local manufacturing facility. Samples of twenty-five are taken every hour and checked for plating cracks or other surface non-conformities. Data for a typical 24-hour production day are shown in Table 10–9.

**TABLE 10–9**

Steel Rod Plating Nonconformity Data

| Sample No. | Sample Size | No. of Nonconformities |
|------------|-------------|------------------------|
| 1 | 25 | 2 |
| 2 | 25 | 0 |
| 3 | 25 | 1 |
| 4 | 25 | 3 |
| 5 | 25 | 5 |
| 6 | 25 | 5 |
| 7 | 25 | 6 |
| 8 | 25 | 4 |
| 9 | 25 | 2 |
| 10 | 25 | 0 |
| 11 | 25 | 1 |
| 12 | 25 | 0 |
| 13 | 25 | 0 |
| 14 | 25 | 2 |
| 15 | 25 | 5 |
| 16 | 25 | 0 |
| 17 | 25 | 0 |
| 18 | 25 | 3 |
| 19 | 25 | 0 |
| 20 | 25 | 0 |
| 21 | 25 | 4 |
| 22 | 25 | 0 |
| 23 | 25 | 1 |
| 24 | 25 | 1 |

a. Construct an np chart from the data in Table 10–9.

b. Interpret the control chart with respect to any patterns discussed in this chapter or in Chapter 9.

6. For the np chart in problem 5, what are the boundaries for:
   a. Zone A above the centerline
   b. Zone B above the centerline
   c. Zone C above the centerline
   d. Zone A below the centerline
   e. Zone B below the centerline
   f. Zone C below the centerline

7. Circuit cards are produced on an automated assembly line. Every tenth circuit card is checked to ensure that all welds on the card are good. Data from the first 300-card lot (of which thirty are tested) produced for the day are shown in Table 10–10.

**TABLE 10–10**
Circuit Card Nonconformity Weld Data

| Card No. | No. Bad Welds |
| --- | --- |
| 1 | 0 |
| 2 | 4 |
| 3 | 0 |
| 4 | 1 |
| 5 | 2 |
| 6 | 1 |
| 7 | 0 |
| 8 | 0 |
| 9 | 0 |
| 10 | 1 |
| 11 | 1 |
| 12 | 0 |
| 13 | 2 |
| 14 | 3 |
| 15 | 5 |
| 16 | 4 |
| 17 | 5 |

**TABLE 10–10** *(cont'd.)*

| Card No. | No. Bad Welds |
|----------|---------------|
| 18 | 3 |
| 19 | 1 |
| 20 | 0 |
| 21 | 0 |
| 22 | 1 |
| 23 | 0 |
| 24 | 0 |
| 25 | 0 |
| 26 | 0 |
| 27 | 2 |
| 28 | 0 |
| 29 | 1 |
| 30 | 0 |

   a. Construct the c chart from the data in Table 10–10.

   b. Interpret the control chart with respect to control chart patterns discussed in this chapter or Chapter 9.

8. For the c chart in problem 7, what are the boundaries for:
   a. Zone A above the centerline
   b. Zone B above the centerline
   c. Zone C above the centerline
   d. Zone A below the centerline
   e. Zone B below the centerline
   f. Zone C below the centerline

9. The leather that is used for handmade horse saddles and bridles of all sizes and varieties arrives at the plant on huge rolls. Because the high-grade leather is very expensive and the end sales price to the consumer of handmade products is very high, the last thing the company wants is to sell saddles and bridles that have blemishes in the leather to consumers. This would definitely result in dissatisfied customers who would probably return the blemished products and not purchase any future products. u Charts are constructed for the rolls of leather to monitor the quality of the leather that is received. The data for a typical roll are shown in Table 10–11.

**TABLE 10–11**
Saddle Leather Nonconformity Data

| Sample No. | Square Feet of Leather | No. Nonconformities |
| --- | --- | --- |
| 1 | 6.2 | 8 |
| 2 | 5.8 | 6 |
| 3 | 4.5 | 5 |
| 4 | 7.0 | 3 |
| 5 | 6.4 | 2 |
| 6 | 4.5 | 3 |
| 7 | 6.0 | 4 |
| 8 | 9.2 | 1 |
| 9 | 7.5 | 2 |
| 10 | 3.5 | 0 |
| 11 | 2.0 | 0 |
| 12 | 5.1 | 1 |
| 13 | 6.8 | 3 |
| 14 | 5.4 | 5 |
| 15 | 4.5 | 0 |
| 16 | 3.5 | 1 |
| 17 | 4.6 | 2 |
| 18 | 3.4 | 3 |
| 19 | 2.5 | 2 |
| 20 | 1.5 | 0 |

    a. Construct the u chart from the data in Table 10–11.

    b. Interpret the u chart with respect to control chart patterns discussed in this chapter and Chapter 9.

10. For the u chart in problem 9, what are the boundaries for:
    a. Zone A above the centerline
    b. Zone B above the centerline
    c. Zone C above the centerline
    d. Zone A below the centerline
    e. Zone B below the centerline
    f. Zone C below the centerline

# 11

# Step 5: Process Capability and Process Capability Indices

## 11.1 Chapter Objectives

After completing this chapter, the student should be able to:

- Understand the theory behind inherent capability, process capability, and process capability indices.
- Complete the calculations for inherent capability, process capability, and process capability indices.
- Interpret what inherent capability, process capability, and process capability indices say with respect to the process.
- Determine the process capability indices of Cp and Cpk and interpret what the indices say with respect to the process.
- Understand the differences between the Cp index and the Cpk index with respect to the central tendency and variability of the process.
- Document and rebaseline the process using the process capability indices of Cp and Cpk and interpret the change of process capability indices in terms of process improvement.

## 11.2 Key Terms

| | | |
|---|---|---|
| Cp | Inherent Capability | Process Capability Indices |
| Cpk | Process Capability | |

## 11.3 Introduction

It should be evident by now that variation exists in processes. And while variation will always exist in nature, it is the enemy of quality. This chapter looks at further quantification of variation as it relates to customer specifications. In other words, it is desirable to know if the process is capable of delivering the product or service with respect to the requirement. Knowing the process capability will help determine the percent of product scrap or rework generated from the process. The roadmap steps to date and the detail of Step 5 are shown in Figure 11–1.

## 11.4 Step 5: Process Capability

**Process capability** is a basic step in any quality program. It is used to analyze the extent to which the product or service produced by the process conforms to the customer specification. It also gives an estimate of the highest quality level that can possibly be achieved by the process. Process capability can be determined at the beginning of the project through a pilot or preliminary study, or it can be determined and monitored on a continuing basis during manufacturing.

### 11.4.1 Step 5A: Inherent Capability

The first step to process capability is **inherent capability**, or the range of variation that will include almost all of the product coming out of the manufacturing process when no assignable variation is present. The only variation present in inherent capability is common cause variation. Inherent capability can be thought of as depicting what the current process is producing. Although in some texts process capability is sometimes used interchangeably with inherent capability, this text uses the strictest definition in that it is only after customer requirements or specifications have been applied with respect to the inherent capability that the transformation to process capability takes place. It is also important to restate the fact that because variation is an important part of the quality definition, it is an important part of both inherent and process capabilities as well.

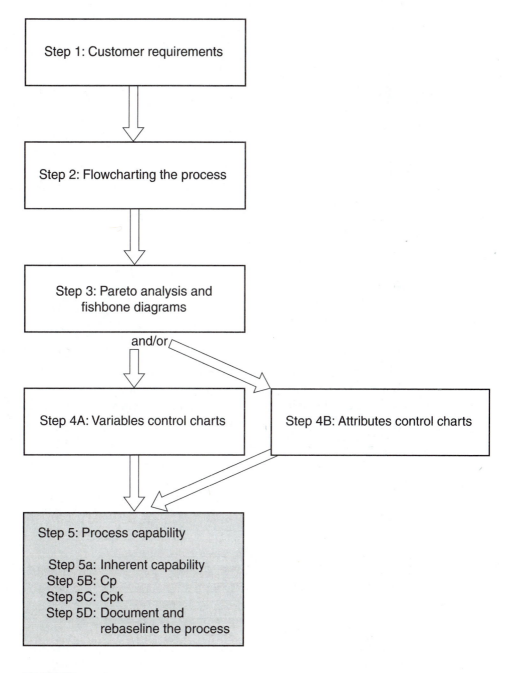

**FIGURE 11–1**
Roadmap Step 5

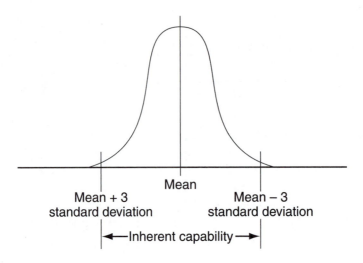

**FIGURE 11–2**
Inherent Capability

When the definition of inherent capability states the phrase "almost all," this implies a mathematical consideration. Remember from Chapter 3 that the mean ± 3 standard deviations includes 99.728% of the population, or almost all of the population. So, when considering inherent capability, the range of variation that will encompass "almost all" of the process is the mean ± 3 standard deviations, or the 6 sigma spread of the process. Graphically, the inherent capability of a process is shown in Figure 11–2.

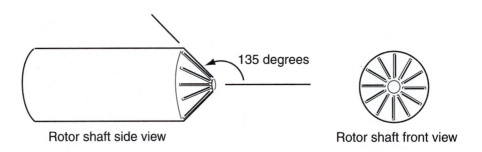

**FIGURE 11–3**
Rotor Shaft Drawing

**EXAMPLE 1**  An automated process for making rotor shafts for electric motors has been targeted by the Industrial Engineering Department as a candidate for an inherent capability study. The rotor shaft end that the individual rotor laminations fit on is angled at 135 degrees, and it is this angle that is the focus of the study because the angle is a critical parameter. A diagram of the rotor shaft is shown in Figure 11–3.

The engineers decide to sample twenty-five rotor shafts and measure the angle to determine the inherent capability. The sample data, measured in degrees, were collected and are shown in Table 11–1.

**TABLE 11–1**
Data for Rotor Shaft Angle (in Degrees)

| | | | | |
|---|---|---|---|---|
| 135.0 | 134.9 | 135.0 | 135.0 | 135.1 |
| 135.0 | 134.8 | 135.1 | 134.9 | 134.9 |
| 135.3 | 134.7 | 135.2 | 135.1 | 134.9 |
| 135.2 | 135.0 | 134.9 | 135.1 | 134.8 |
| 135.1 | 135.0 | 135.0 | 135.0 | 135.0 |

The process mean is 135.0 degrees and the standard deviation is 0.14 degrees. The inherent capability is the area of the mean ± 3 standard deviations, which is from 134.58 degrees to 135.42 degrees. The inherent capability for the rotor shaft process is shown in Figure 11–4.

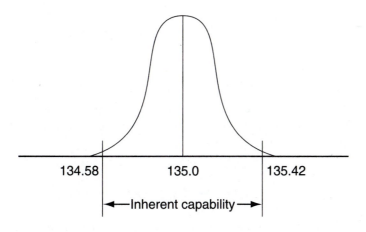

134.58          135.0          135.42

←—Inherent capability—→

**FIGURE 11–4**
Inherent Capability of Rotor Shaft Process (in Degrees)

## 11.4.2 Step 5B: Cp

**Process capability indices** are a technique for specifying quality by statistical measures. They are another way to quantify the degree to which the process is or is not meeting the customer requirement. The first process capability index to look at is **Cp**. The Cp index is strictly a variation index that compares the specification width to the process width. The formula for Cp when a two-sided specification limit is present is:

$$Cp = \frac{|USL - LSL|}{6\sigma}$$

where:   USL = the upper specification limit
         LSL = the lower specification limit
         $6\sigma$ = the process width (mean $\pm\, 3\sigma$)

For a specification limit in which there is only one side (for example, a specification with either the upper specification or lower specification), the formula for the Cp index becomes:

$$Cp = \frac{|USL - \mu|}{3\sigma} \quad \text{for upper specification}$$

. . . or . . .

$$Cp = \frac{|\mu - LSL|}{3\sigma} \quad \text{for lower specification}$$

---

**EXAMPLE 1**   A process has a width of 40 (the 6 sigma spread of the mean ± 3 standard deviations). The specification width is 30 (USL – LSL). The process is depicted in Figure 11–5 with the calculation of the Cp index following.

$$Cp = \frac{30}{40} = 0.75$$

Notice that with a Cp value less than 1.0, some products will be produced that will lie outside of the specification width. This will result in rework unless the process is improved and the variability is reduced. The value of 0.75 will correspond to some percentage that lies outside of the specification width. It is important to remember that each numeric value of Cp will correspond to a specific percentage that will not lie

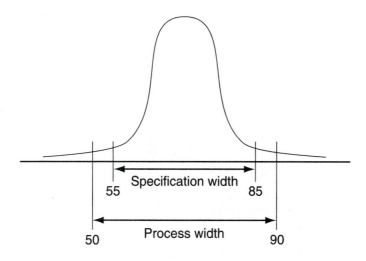

**FIGURE 11–5**
Process Width and Specification Width Example for Cp = 0.75

within the specification limits. The higher the value of Cp, the lower the percent of scrap and rework will be.

---

EXAMPLE 2   A variation reduction program is initiated and the process width is now 30. The improved process is shown in Figure 11–6 and the calculation of the new Cp follows.

$$Cp = \frac{30}{30} = 1.00$$

Now that the variability has been reduced, the Cp value becomes 1.00, which means that the process spread (6 sigma) is equal to the specification spread.

---

An interesting link can now be formed to understand the numeric value of the Cp index: the process spread (6 sigma) represents 99.728% and is equal to the specification width. This means that 99.728% of the products produced will fall within the specification limit, or 0.272% will lie outside of the specification limits, which will result in scrap or rework of 0.272% in its current state. This can now be converted to parts per million (PPM) scrap. In this case, 2720 PPM can be expected from the

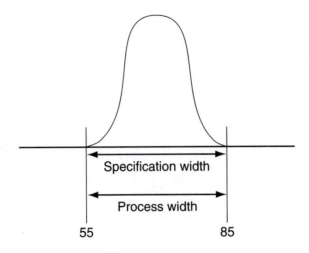

**FIGURE 11–6**
Process Width and Specification Width Example for Cp = 1.00

process as it currently stands. The 2720 PPM figure is obtained from the logic shown below, in which the decimal is shifted one position to the right until the PPM can be determined.

- 99.728 out of 100 units will be good and 0.272 units will be scrap
- 997.28 out of 1000 units will be good and 2.72 units will be scrap
- 9972 out of 10,000 units will be good and 27.2 units will be scrap
- 99,720 out of 100,000 units will be good and 272 units will be scrap
- 997,200 out of 1,000,000 units will be good and 2720 units will be bad

The reader should now begin to see the value of process capability and the process capability index of Cp. It yields a value that can be translated into a percentage of scrap or rework that results from process variation.

---

**EXAMPLE 3** Process improvement continues and the process spread is now 20. This improved process is shown in Figure 11–7, with the calculation for Cp following.

$$Cp = \frac{30}{20} = 1.67$$

The process capability index is now above 1.00 and as variability reduction continues, the amount of scrap and rework will continue to diminish.

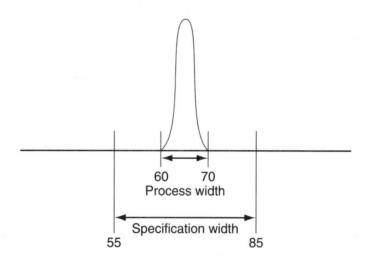

**FIGURE 11–7**
Process Width and Specification Width Example for Cp = 1.67

Several things should become apparent with respect to Cp:

- It is a powerful technique for relating process variation to specification width.
- The higher the numeric value of Cp, the better.
- It quantifies the amount of scrap and rework that result from a process.
- The mean of the process must be equal to the mean of the specification for Cp to be useful. To prove this look at Figure 11–8, where the process spread is equal to the specification spread. Although Cp will be 1.00, it can be seen that the process will produce virtually nothing that will lie within the specification limits. Thus, caution must be exercised when using Cp to ensure that the mean of the process is at the center of the specification, especially when considering process improvement or determining the percent of scrap or rework generated from the process. The calculation of Cp for this example is:

$$Cp = \frac{30}{30} = 1.00$$

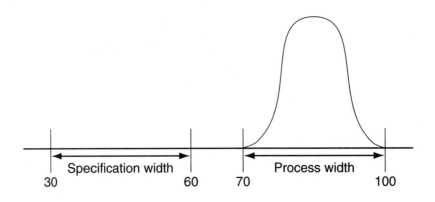

**FIGURE 11–8**
Example of Importance of Centering Process Width to Specification Width When
Using Cp

## 11.4.3 Step 5C: Cpk

The subject of centering leads to a more useful process capability index: **Cpk**.
Whereas Cp considers only variation, Cpk considers both the central tendency of the
process in relation to the specifications as well as variability. The value of considering
both central tendency and variation is that the approach to reducing variability is different than the approach to centering, as is the level of difficulty associated with each.
Because Cpk considers both central tendency and variability, it gives better insight
into which of these, if not both, is the largest contributor to product non-quality, and
which one should be tackled first. In essence, Cpk considers the totality of quality as
the definition of quality considers both central tendency and variability. Cpk is
defined below:

$$\text{Cpk} = \text{Minimum}\left\{\frac{\text{USL} - \mu}{3\sigma} \text{ or } \frac{\mu - \text{LSL}}{3\sigma}\right\}$$

When using Cpk, as opposed to Cp, it is not necessary to ensure that the mean of
the process is at the center of the specification width. In the Cpk formula, this concern
is taken care of with the introduction of the centering aspect. Similar to Cp, when
considering the numeric value of Cpk, the higher the number, the better.

---

**EXAMPLE 1**    A process possesses an average of 20 with a process width of 20. The USL is 25 and the
LSL is 15. The picture is shown in Figure 11–9 and calculation for Cpk follows.

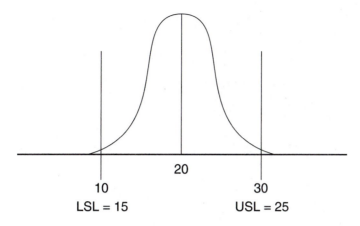

**FIGURE 11–9**
Process Example for Cpk = 0.5

$$Cpk = Minimum\left\{\frac{25-20}{10} \text{ or } \frac{20-15}{10}\right\}$$
$$Cpk = Minimum\{0.5 \text{ or } 0.5\}$$
$$Cpk = 0.5$$

It can be seen that the numerical value of Cpk lies under 1.0, which means that the process will produce scrap. This is verified by looking at the graph and observing that the tail ends of the distribution lie outside of the specification limits. Thus, variability reduction needs to occur.

---

**EXAMPLE 2**    Process improvement takes place and the improved process is shown in Figure 11–10. Note that with the new process both variability reduction and process centering shifting have occurred.

$$Cpk = Minimum\left\{\frac{USL-\mu}{3\sigma} \text{ or } \frac{\mu-LSL}{3\sigma}\right\}$$
$$Cpk = Minimum\left\{\frac{25-18}{2} \text{ or } \frac{18-15}{2}\right\}$$
$$Cpk = Minimum\{3.5 \text{ or } 1.5\}$$
$$Cpk = 1.5$$

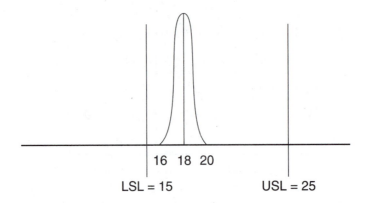

**FIGURE 11–10**
Process Example for Cpk = 1.5

The Cpk value for the new process is 1.5, which is an improvement, but although the variability reduction looks good, the centering of the process has definitely drifted with respect to the specification.

EXAMPLE 3   Once again, a process improvement initiative takes place with the focus being on the centering of the process, and the results are shown in Figure 11–11.

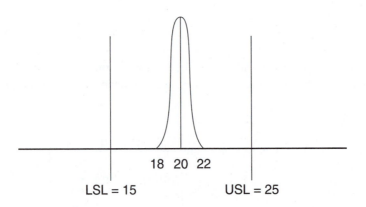

**FIGURE 11–11**
Process Example for Cpk = 2.5

$$Cpk = \text{Minimum}\left\{\frac{USL - \mu}{3\sigma} \text{ or } \frac{\mu - LSL}{3\sigma}\right\}$$

$$Cpk = \text{Minimum}\left\{\frac{25 - 20}{2} \text{ or } \frac{20 - 15}{2}\right\}$$

$$Cpk = \text{Minimum}\{2.5 \text{ or } 2.5\}$$

$$Cpk = 2.5$$

Notice that the numeric value of the Cpk is increasing. Also notice that from the graph the process has been centered. It is important to remember that although a picture of the process is not necessary, it is highly recommended to assist in viewing the situation. The Cpk improved to 1.5, but the graph clearly showed that the process was improved with respect to centering.

When looking at some guidelines for values of Cpk, it is practical to strive for a minimum value of 1.33. A value of 2.00 is more desirable, and an excellent Cpk value would be 5.00 or greater.

# 11.5  A Side-by-Side Look at Cp and Cpk

This is an excellent time to show the Cp and Cpk indices side by side in an example to demonstrate the differences between the two, and the importance of looking at both of them together.

EXAMPLE 1    Recall the process depicted in Example 1 in Section 11.4.3. The process is shown in Figure 11–12 with the process capability indices indicated in the figure caption.

In the preceding process both the Cp and Cpk values are equal, due to the fact that the process was centered at the midpoint of the specification. In continuing, recall that the process was improved in Example 2 of Section 11.4.3, but that the improvement was in variability reduction while the mean of the process shifted to the left, as shown in Figure 11–13. The process capability indices values are indicated in the figure caption.

Now, the value of both Cp and Cpk have increased but they are no longer equal. In fact, looking at the Cp index by itself would not be beneficial at all because unless the process is centered at the midpoint of the specification limit the value of Cp is somewhat misleading. The next improvement Example 3 of Section 11.4.3 focused around shifting the mean of the process to reside at the midpoint of the specification limit. This is shown in Figure 11–14, again with the values of the process capability indices indicated in the figure caption.

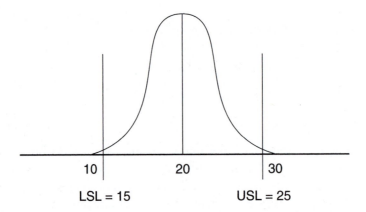

**FIGURE 11–12**
Process Example for Cp = 0.5 and Cpk = 0.5

In this third iteration, it can be seen that the Cpk value increased (due to centering) while the Cp value remained the same. Cpk is responsive to both changes in central tendency of the process and variability while Cp is just responsive to variability. It is recommended to look at both the Cp and Cpk values side by side to get a total picture of what is happening within the process.

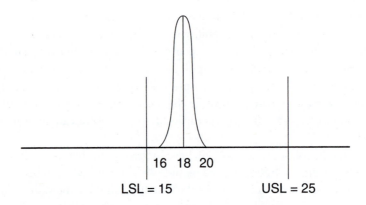

**FIGURE 11–13**
Process Example for Cp = 2.5 and Cpk = 1.5

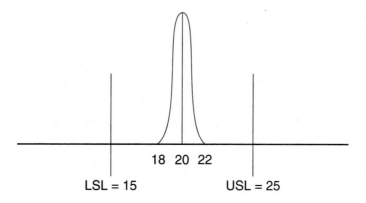

**FIGURE 11–14**
Process Example for Cp = 2.5 and Cpk = 2.5

## 11.6 The Engineering Specification Pitfall

It is important to take a moment and mention a line of thinking that is sometimes used when it comes to process capability, but is definitely not endorsed: changing the engineering specification in order to make the process capable. In other words, there are two ways to increase the value of the process capability indices of Cp and Cpk:

1. Implement an effective process improvement initiative that addresses both variability reduction and central tendency.
2. Increase the engineering specification.

It is true that widening the specification width and doing nothing to reduce the variability of the process will indeed increase the process capability, but it is definitely not a correct quality philosophy. Arbitrary extension of the engineering specification can have a detrimental impact on product performance. This topic is addressed to point out a definite pitfall to avoid when considering process capability and capability indices.

## 11.7 Step 5D: Document and Rebaseline the Process

The importance of baselining is highlighted in this chapter with the examples of both Cp and Cpk. Notice that in each of the series of examples of Cp and Cpk, continual process improvement occurred, but without documentation one would not know how much improvement resulted. Every time that process is improved, documentation and rebaselining should occur, both with respect to central tendency and variability. One very basic reason for documentation and rebaselining is that as the process is improved, the resulting product will be improved. And as the product is improved, new customers and new markets can be obtained. As the Cp and Cpk numerically increase, reduced parts per million scrap will occur, which means that the product quality is improved.

An example of documentation and rebaselining is demonstrated next.

---

**EXAMPLE 1**   A company produces stampings. One of the critical customer requirements is length, which possesses a length of 5.0 in. ± .1 in. The current stamping process data gathered on January 1st yields the following information:

- Process average = 5.0 in.
- Range = 2.0 in.
- Average + 3 standard deviations = 6.0 in.
- Average − 3 standard deviations = 4.0 in.
- Cp = 0.1
- Cpk = 0.1

An intensive process improvement initiative is undertaken for the next two months and all of the process improvements have been appropriately documented. On April 1st data are collected to rebaseline the process:

- Process average = 5.0 in.
- Range = 2.0 in.
- Average + 3 standard deviations = 5.1 in.
- Average − 3 standard deviations = 4.9 in.
- Cp = 1.0
- Cpk = 1.0

Imagine the reaction of the customer when she sees the quantifiable data depicting the company's first process improvement initiative? Even if the customer was initially dismayed as to the performance of the stamping company on January 1st, one would be hard pressed to imagine that the customer would abandon her contract with the stamping company now, especially if process improvement continues into further stages. In continuing the example, suppose the customer is still pushing for improved quality and the stamping company aggressively continues its process improvement initiative over the next two months, documenting all of the improvements in exhaustive detail. On July 1st the company once again rebaselines and the process data are detailed below.

- Process average = 5.0 in.
- Range = 0.1 in.
- Average + 3 standard deviations = 5.05 in.
- Average − 3 standard deviations = 4.95 in.
- Cp = 2.0
- Cpk = 2.0

As the initiative continues, it becomes very obvious of the importance of documentation and rebaselining. In addition to the reasons spelled out above in this example, the documentation also provides for an audit trail and historical accounting of the process improvement, which is instrumental when applying for ISO certification.

## 11.8 Summary

- Inherent capability is the range of variation that includes almost all of the product as it comes out of the manufacturing process. "Almost all" is defined mathematically as the area that lies between the average (mean) ± 3 standard deviations, which constitutes 99.728% of the variation.
- Process capability is the extent to which the product or service produced by the process conforms to the customer specification.
- The process capability index Cp is a ratio of the tolerance width to the process width.
- The mean of the process must be centered at the midpoint of the specification width for the Cp index to be useful.
- The higher the Cp index value, the more capable the process is of meeting the customer requirements.
- A Cp value of less than 1.0 denotes an incapable process; a Cp value equal to 1.0 denotes a marginal process; and Cp values that continue to increase beyond 1.0 indicate a process that is becoming more and more capable.
- Cp values are used to indicate the amount of scrap and rework in parts per million.
- The process capability index of Cpk considers both the centering of the process as well as the variability of the process with respect to the customer requirement.
- The higher the Cpk value, the more capable the process is of meeting the customer requirement.
- A good target value for Cpk is 1.33; a more desirable Cpk value is 2.00; and an excellent Cpk value is 5.00 or greater.
- It is important to document the process and rebaseline to quantifiably show the improvement of the process.

## Review Questions

1. Process capability is:
   a. the width of the specification limit divided by the process width
   b. the range of variation of the process that includes "almost all" of the variation
   c. the extent to which the product or service produced by the process can conform to the customer specification
   d. none of the above

2. The numerator of the Cp index is:
   a. the specification width
   b. the process width
   c. the midpoint of the specification width
   d. none of the above

3. The process capability index that considers both central tendency as well as variability is:
   a. Cp
   b. Cpk
   c. both indices
   d. neither index

4. Inherent capability is:
   a. the width of the specification limit divided by the process width
   b. the range of variation of the process that includes "almost all" of the variation
   c. the extent to which the product or service produced by the process can conform to the customer specification
   d. none of the above

5. A good initial target value for Cpk is:
   a. 1.33
   b. 2.00
   c. 5.00
   d. for an initial target, any value of Cpk is good
   e. none of the above

6. A marginal process has:
   a. a Cp value of 2.00
   b. a Cpk value of 2.00
   c. a Cp value of 0.90
   d. a Cpk value of 1.33
   e. none of the above

7. The Cp denominator for a one-sided specification limit is:
   a. 6 standard deviations
   b. 3 standard deviations
   c. 2 standard deviations
   d. none of the above

8. The denominator of Cpk is:
   a. the process mean + 1 standard deviation
   b. the process mean + 2 standard deviations
   c. the process mean + 3 standard deviations
   d. none of the above

9. The process capability index that considers only variation is:
   a. Cp
   b. Cpk
   c. both indices
   d. neither index

10. Keeping track of the progress during process improvement for the purposes of audit trails is called:
    a. rebaselining
    b. documentation
    c. inherent capability
    d. process capability
    e. none of the above

11. The numerator of the Cpk index is:
    a. the variability of the process
    b. a difference of one of the specification limits with respect to the process mean
    c. the difference between the upper and lower specification limits
    d. none of the above

12. The denominator of the Cp index is:
    a. the mean of the process
    b. difference of one of the specification limits with respect to the process mean
    c. the difference between the upper and lower customer requirements
    d. none of the above

13. Both Cp and Cpk adhere to a simple rule.
    a. The lower the value, the better the process performs.
    b. The higher the value, the better the process performs.
    c. Cp and Cpk will always be equal.
    d. Nothing can be determined until the actual value is computed.

14. Of the following, which would be the best situation for minimum variation?
    a. The Cp value is >5.00.
    b. The Cpk value is >5.00.
    c. Both the Cp and Cpk values are >5.00.
    d. None of the above.

15. When the process is centered at the midpoint of the specification limit:
    a. Cp will be greater than Cpk
    b. Cp will be less than Cpk
    c. Cp will be equal to Cpk
    d. none of the above

16. An incapable process has:
    a. a Cpk value = 3.00
    b. a Cp value = 1.33
    c. a Cpk value =1.33
    d. none of the above

17. An excellent Cpk value is:
    a. 5.00
    b. 6.00
    c. 7.00
    d. all of the above
    e. none of the above

18. If the mean of the process does not equal the midpoint of the specification limits:
    a. Cp will equal Cpk
    b. Cp will not equal Cpk
    c. no statement regarding the relationship of Cp to Cpk can be made
    d. none of the above

19. An unacceptable method for improving process capability is:
    a. arbitrarily expanding the specification
    b. starting a process improvement initiative
    c. both a and b
    d. none of the above

20. "Almost all" is mathematically defined as:
    a. the area under the normal curve represented by the mean ± 1 standard deviation
    b. the area under the normal curve represented by the mean ± 2 standard deviations
    c. the area under the normal curve represented by the mean ± 3 standard deviations
    d. none of the above

# Problems

1. A process is being studied to determine the process capability. Data have been collected from the process over the last month. The data are shown in Table 11–2, and the measurement unit is centimeters.

**TABLE 11–2**
Process Data

| | | | |
|------|------|------|------|
| 45.2 | 46.1 | 46.4 | 46.4 |
| 44.7 | 46.4 | 45.2 | 45.5 |
| 46.3 | 46.4 | 46.2 | 47.1 |
| 45.4 | 45.9 | 46.4 | 46.8 |
| 46.7 | 46.7 | 46.7 | 46.8 |
| 46.9 | 45.2 | 45.3 | 45.6 |
| 46.5 | 46.4 | 46.5 | 46.4 |
| 45.7 | 46.4 | 47.0 | 45.5 |
| 46.5 | 46.4 | 46.4 | 46.3 |
| 46.4 | 46.6 | 46.3 | 46.4 |
| 46.1 | 46.6 | 46.4 | 46.9 |
| 45.8 | 45.9 | 46.2 | 46.3 |
| 46.4 | 46.4 | 46.5 | 46.6 |
| 46.7 | 46.4 | 45.7 | 46.4 |
| 46.4 | 46.5 | 46.6 | 46.6 |
| 46.4 | 46.4 | 46.4 | 46.5 |

a. Construct the histogram for the data in the table.
b. Determine the mean of the process.
c. What is the inherent capability of the process?

2. A customer requirement has a lower specification limit of 45.6 cm and an upper specification limit of 46.7 cm.
a. Using the data in Table 11–2, calculate Cp.
b. Using the data in Table 11–2, calculate Cpk.
c. What conclusions, if any, can be drawn from the process capability indices?

3. A new customer requirement has been received. The lower specification limit is now 46.3 cm and the upper specification limit is 47.7 cm.
a. Using the data in Table 11–2, calculate Cp.
b. Using the data in Table 11–2, calculate Cpk.
c. What conclusions, if any, can be drawn from the process capability indices?
d. What conclusions, if any, can be drawn by comparing the process capability indices in problem 2 to the ones in this problem?

4. Draw the diagram for each of the following situations.
   a. Cp = 3.00
   b. Cp = 0.67
   c. Cp = 2.00
   d. Cp = 1.50

5. Draw the diagram for each of the following situations.
   a. Cpk = 5.00
   b. Cpk = 2.00
   c. Cpk = 0.50
   d. Cpk = 1.50

6. Fill in the blanks for each of the following statements.
   a. When the upper and lower specification limits are equal to the mean ± 2 standard deviations, then the process will result in _% scrap or rework.
   b. When the upper and lower specification limits are equal to the mean ± 4 standard deviations, then the process will result in _% scrap or rework.
   c. When the upper and lower specification limits are equal to the mean ± 5 standard deviations, then the process will result in _% scrap or rework.
   d. When the upper and lower specification limits are equal to the mean ± 6 standard deviations, then the process will result in _% scrap or rework.

7. The data in Table 11–3 are obtained by life cycle testing a random sample of products. The data, measured in cycles before failure, will be provided to a process improvement team, of which you have been selected to lead.

**TABLE 11–3**
Product Life Cycle Data

| | | | |
|-----|-----|-----|-----|
| 600 | 599 | 600 | 601 |
| 602 | 601 | 594 | 599 |
| 599 | 603 | 601 | 600 |
| 597 | 600 | 602 | 601 |
| 603 | 602 | 599 | 598 |
| 599 | 598 | 600 | 602 |
| 601 | 599 | 601 | 597 |
| 598 | 601 | 600 | 602 |
| 599 | 601 | 599 | 600 |
| 601 | 597 | 600 | 599 |
| 592 | 600 | 603 | 600 |

  a. Draw the histogram for the data in Table 11–3.
  b. What is the process mean?
  c. What is the inherent capability of the process?

8. The lower specification of the customer requirements is 595 cycles and the upper specification is 606.
  a. Using the data in Table 11–3, compute Cp.
  b. Using the data in Table 11–3, compute Cpk.
  c. What conclusions, if any, can be drawn from looking at the process capability indices?

9. A process improvement initiative has been started on the process described in problem 7, and one of the key improvements has been changing the material of the product to a relatively new, stronger, and lighter polymer. There is concern in the Engineering Department, however, that the interaction of the increased strength to the decreased weight may result in the product not improving as much as had been perceived by other departments in the organization. You, as lead of the team, decide to gather data to determine the impact of the new material on product performance. You direct that data be gathered again from the process. The data, measured in cycles before failure, were collected and are shown in Table 11–4. The customer requirement is specified in Problem 8.

**TABLE 11–4**
Product Life Cycle Data (After Improvement Initiative)

| | | | |
|-----|-----|-----|-----|
| 599 | 598 | 599 | 600 |
| 601 | 600 | 593 | 598 |
| 598 | 602 | 600 | 599 |
| 596 | 599 | 601 | 600 |
| 602 | 601 | 598 | 597 |
| 598 | 597 | 599 | 601 |
| 600 | 598 | 600 | 596 |
| 597 | 600 | 599 | 601 |
| 598 | 600 | 598 | 599 |
| 600 | 596 | 599 | 598 |
| 591 | 599 | 602 | 599 |

  a. Draw the histogram of the process after the improvement has been implemented.
  b. What is the new process mean?

    c. What is the inherent capability?

    d. Compute Cp.

    e. Compute Cpk.

10. What conclusions can be made regarding the newly improved process versus the initial process? Note: Make sure your conclusions are supported by quantitative data. Be sure to address the change, if any, of the Cp and Cpk values and what the change means in terms of process improvement.

# 12

## Step 6: Design of Experiments (DOE)

## 12.1  Chapter Objectives

After completing this chapter, the student should be able to:

- Understand what is meant by design of experiments.
- Tell the difference between trial and error experiments, one at a time experiments, full factorial experiments, and fractional factorial experiments.
- Set up a basic experiment.
- Conduct a basic experiment.
- Interpret the results of a basic experiment.
- Develop the settings and predict the maximum output of an experiment.

## 12.2  Key Terms

Design Arrays
Design of Experiments
Fractional Factorial
    Experiments

Full Factorial Experiment
Noise
One Factor at a Time
    Experiment

Robust Design
Dr. Genichi Taguchi
Trial and Error
    Experiment

## 12.3  Introduction

**Design of experiments** is the last step in the roadmap to process improvement and improved quality. Design of experiments can be implemented at any stage of the product or process life cycle (for example, cradle to grave), but it is shown as Step 6 in the roadmap. This is because it is extremely important to understand the process, which Steps 1 through 5 systematically endorse, before jumping into an experimental design. As seen in the roadmap, the optimization of an experiment hinges on process understanding, inclusive of the philosophies of techniques such as the Pareto and Fishbone diagrams, which are part of the roadmap steps. The roadmap steps to date, inclusive of Step 6, are shown in Figure 12–1.

## 12.4  What Is Design of Experiments?

The best way to start discussion on Design of Experiments (DOE) is to realize that DOE is something that is used every day by people in government, industry, chefs, coaches of team sporting events, and many more. People may not realize it, but they conduct experimental designs in a similar manner to the way it is done in the laboratory. People vary different factors that interact to try to come up with some winning combination. The manager of a baseball, football, soccer, or basketball team will juggle the line-up in order to find out the combination that will produce a winning team. A chef will experiment with different ingredients in different amounts in an attempt to come up with a winning recipe for his already highly acclaimed meatloaf. Supervisors may change the workload between people, or change the methodology from manual calculations to a computer-based program in an effort to increase efficiency or increase productivity. Agriculture labs conduct experiments altering the types of fertilizers, water, and other factors in an effort to increase yields. All of these examples are similar in that different levels of ingredients are used to achieve some desired outcome, or quality characteristic. And in essence, that is what Design of Experiments is about. DOE is involved with the manipulating of controllable factors in order to not only determine their effect on some desired outcome, but to find the optimum combination of factors that will yield the maximum desired outcome.

Recall the diagram of a process from Chapter 3 (reshown in Figure 12–2), where all of the factors of materials, manpower, machinery, methods, and others are shown as dials on the process box. This pictorial depicts what DOE is all about: determining the optimum settings or levels of the resources to produce the product that meets the quality characteristics. It must be stated here that it is not the intentions of the author to delve into all of the aspects of Design of Experiments, such as inner and outer arrays, orthogonality, and noise, just to name a few. That requires an entire course in itself. Rather, the intent is to demonstrate the application and role experiments play in process improvement.

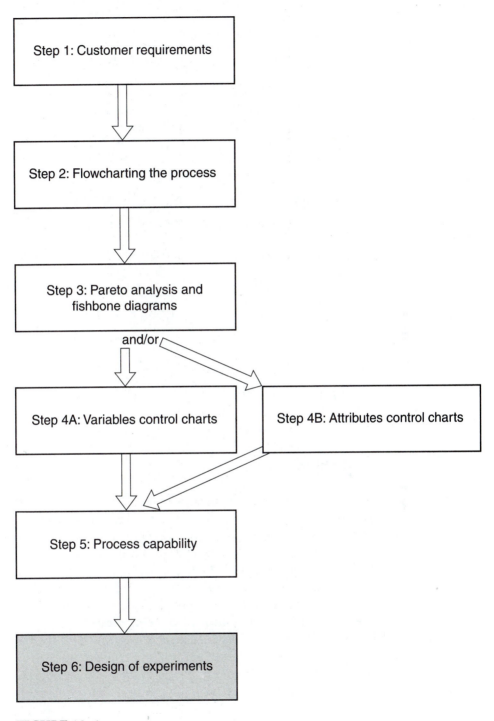

**FIGURE 12–1**
Roadmap Step 6

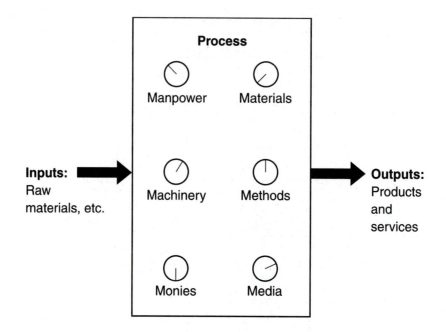

**FIGURE 12–2**
Process Diagram

## 12.5 Types of Experiments

In looking at any one of the situations in Section 12.4, it can be seen that it would be illogical just to change every single factor by a different amount at the same time. Imagine what would happen if the chef changed all of the ingredients in different amounts all at the same time. The result would be total chaos. The chef would have no idea as to which ingredient in which amount and what interaction of ingredients contributed the most to the improvement to his recipe. Thus, implied in Design of Experiments is that there is some logical approach to gain information about the quantity and type of controllable factors involved in the experiment.

There are different types of experimentation:

- **trial and error**
- **one factor at a time**
- **full factorial**
- **fractional factorial**

### 12.5.1 Trial and Error Experiments

Trial and error experimentation is where one factor is manipulated without regard to the other factors.

---

**EXAMPLE 1**    In the chef example cited in Section 12.4, assume that the chef may have some particular insight that one factor, say salt, would improve the taste of his meatloaf. The chef may change that one ingredient and make a meatloaf each time the type or amount of salt is changed.

---

**EXAMPLE 2**    A particular product consists of an alloy that is composed of three different metals. Specifically, the alloy mixture consists of 40% of metal 1, 35% of metal 2, and 25% of metal 3. It is desired by management to increase the overall strength of the alloy that is used in the product. The product engineer has some particular insight that increasing the percent of metal 1 will definitely increase the strength of the product. The engineer can change the percent of metal 1 in the alloy mixture, manufacture some sample products, and test the overall product strength each time the percent of metal 1 is changed.

---

The downside of this type of experimentation is that in today's environment trial and error may prove to be slow, costly, and inefficient. To assume that corporate world competition is massively employing techniques such as trial and error is incorrect. More times than not, when experiments are conducted, something more sophisticated and efficient is employed. Likewise, trial and error experimentation assumes that everything about the factors and their levels is known, which may or may not necessarily be the case.

### 12.5.2 One Factor at a Time Experiments

This type of experimentation differs from trial and error and is an improvement over trial and error in that:

– A series of factors may be studied but only one factor is adjusted at a time.
– Only one factor is adjusted while all the other factors are held constant.

---

**EXAMPLE 1**    In the chef example, the chef may feel that while there are many ingredients involved in making his meatloaf, there are just a couple of factors that combine to drastically improve the taste of his meatloaf. The ingredients of interest are salt, onions, and

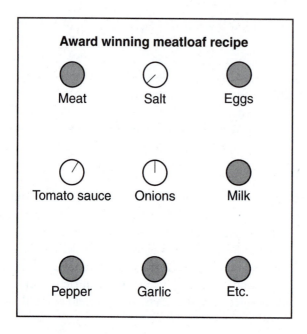

**FIGURE 12–3**
Chef's Process Factors

tomato sauce. Furthermore, he feels that it is only the amount of these ingredients and not the brand of ingredient that needs to be tested. The pictorial for the chef's experiment is shown in Figure 12–3. The factors that will not be adjusted in his experiment are blackened out.

In the first one factor at a time experiment, the salt will be adjusted while the onions and tomato sauce are held constant. In the second experiment, the onions will be adjusted while the salt is returned to its original value and the tomato sauce is held constant. And in the third experiment, the tomato sauce will be adjusted while the salt and onions are returned to their original value and held constant. The experimental layout for this example is as follows:

Original Ingredients and Levels:

Salt:          ½ tsp

Onion:         ½ diced

Tomato Sauce:  ½ cup

| EXPERIMENT | SALT | ONION | TOMATO SAUCE |
|---|---|---|---|
| 1 | **¼ tsp** | ½ diced | ½ cup |
| 2 | ½ tsp | **¾ diced** | ½ cup |
| 3 | ½ tsp | ½ diced | **¾ cup** |

If the chef felt there still needed to be some improvement, then he could continue to perform experiments changing one factor at a time, as shown:

| EXPERIMENT | SALT | ONION | TOMATO SAUCE |
|---|---|---|---|
| 4 | **⅛ tsp** | ½ diced | ½ cup |
| 5 | ½ tsp | **1 diced** | ½ cup |
| 6 | ½ tsp | ½ diced | **1 cup** |
| 7 | **¾ tsp** | ½ diced | ½ cup |
| 8 | ½ tsp | **¼ diced** | ½ cup |
| 9 | ½ tsp | ½ diced | **1¼ cup** |

One factor at a time experimentation is definitely an improvement over trial and error experimentation in that it constitutes a systematic experimental approach. But it is also costly, somewhat inefficient, and there is the high chance that an overestimate of the effect that a single factor has may occur.

## 12.5.3 Full Factorial Experiments

Full factorial experimentation differs from the prior two types in that every possible combination of factors is tested simultaneously at selected levels.

**EXAMPLE 1**    Continuing with the chef example, suppose the chef chooses to again test the three factors affecting the taste of his meatloaf: salt, onions, and tomato sauce. And furthermore, the chef is still convinced that these are the only three factors that can combine in some fashion to dramatically improve the taste of his particular recipe. He is further convinced that the two levels he has chosen to test for each ingredient are the correct levels. The basic setup for his experiment using a full factorial design is:

| FACTORS | LEVEL 1 | LEVEL 2 |
|---|---|---|
| Salt | ⅛ tsp | ¾ tsp |
| Onions | ½ diced | 1½ diced |
| Tomato Sauce | ½ cup thin | 1 cup thin |

The chef's experiment will require eight tests (the three factors at the two levels), and the test combinations are as follows:

| EXPERIMENT | SALT | ONION | TOMATO SAUCE |
|---|---|---|---|
| 1 | Level 1 | Level 1 | Level 1 |
| 2 | Level 1 | Level 1 | Level 2 |
| 3 | Level 1 | Level 2 | Level 1 |
| 4 | Level 1 | Level 2 | Level 2 |
| 5 | Level 2 | Level 1 | Level 1 |
| 6 | Level 2 | Level 1 | Level 2 |
| 7 | Level 2 | Level 2 | Level 1 |
| 8 | Level 2 | Level 2 | Level 2 |

Replacing the levels in the test plan with the specified ingredient levels yields:

| EXPERIMENT | SALT | ONION | TOMATO SAUCE |
|---|---|---|---|
| 1 | ⅛ tsp | ½ diced | ½ cup thin |
| 2 | ⅛ tsp | ½ diced | 1 cup thin |
| 3 | ⅛ tsp | 1½ diced | ½ cup thin |
| 4 | ⅛ tsp | 1½ diced | 1 cup thin |
| 5 | ¾ tsp | ½ diced | ½ cup thin |
| 6 | ¾ tsp | ½ diced | 1 cup thin |
| 7 | ¾ tsp | 1½ diced | ½ cup thin |
| 8 | ¾ tsp | 1½ diced | 1 cup thin |

The full factorial experiment possesses several advantages over the first two types of experiments discussed previously. It is a very thorough form of experimentation with simultaneous testing of all factors at selected levels. From this aspect comes another advantage: it is through this thorough testing of all factors that ensures the conclusions are sound.

Notwithstanding the benefits of full factorial experiments, this type of experimentation carries some very heavy disadvantages. The biggest ones have to do with practicality. The costs associated with time, complexity, and dollar value of the many runs required with full factorials can be very high. Imagine if the chef had further decided to conduct experiments with not only the different amounts of tomato sauce but tomato sauces of different thicknesses as well. The number of experiments would grow to an astronomical number. To prove this, the equation for the number of tests that need to be conducted in a full factorial experiment is as follows:

$$L^f$$

where:  f = the number of factors in the experiment

L = the number of levels for the factors

To get an idea of the size of investment in time, money, and complexity associated with full factorial experimentation, Table 12–1 depicts the number of experiments required with various factors and levels.

**TABLE 12–1**
Full Factorial Number of Runs for Various Factors and Levels

|               | FACTORS | LEVELS | No. of EXPERIMENTS REQUIRED |
|---------------|---------|--------|------------------------------|
| Two Levels:   | 2       | 2      | 4                            |
|               | 3       | 2      | 8                            |
|               | 4       | 2      | 16                           |
|               | 5       | 2      | 32                           |
|               | 6       | 2      | 64                           |
|               | 7       | 2      | 128                          |
|               | 8       | 2      | 256                          |
|               | 9       | 2      | 512                          |
|               | 10      | 2      | 1024                         |
| Three Levels: | 2       | 3      | 9                            |
|               | 3       | 3      | 27                           |
|               | 4       | 3      | 81                           |
|               | 5       | 3      | 243                          |
|               | 6       | 3      | 729                          |
|               | 7       | 3      | 2187                         |
|               | 8       | 3      | 6561                         |
|               | 9       | 3      | 19683                        |
|               | 10      | 3      | 59049                        |
| Others:       | 15      | 2      | 32768                        |
|               | 15      | 3      | 14348907                     |
|               | 32      | 2      | 4294967296                   |

**EXAMPLE 2**   A football coach is thinking about changing the roster of his defensive line to improve effectiveness against the run. His scenario for this mid-season change is simple: the current defensive line has four varsity players and his depth chart for each of the positions is

only one deep with the back-ups all being underclassmen with minimal experience but superior quickness. In addition, the defensive line is performing very sporadically in both practice and games when they change from short cleats (quicker) to medium cleats to long cleats (for digging in). He is at a complete loss because the statistics coming in are yielding very little information as to what to change. He feels that there must exist some combination of varsity experience, freshman speed, and cleat length that will improve his defense and keep his team's chances for postseason playoffs alive. The problem is that he is not comfortable with a trial and error type of experiment. He realizes that he needs to test the roster changes over an entire game and not just one series of plays, or for one quarter. He recalls from his college days that full factorial experiments are very thorough and the results are very accurate, so he considers conducting a full factorial experiment or else his chances of postseason play will quickly diminish. His test plan follows:

| FACTOR | LEVEL 1 | LEVEL 2 | LEVEL 3 |
|---|---|---|---|
| PLAYERS | four seniors | two seniors/two underclassmen | four underclassmen |
| CLEATS | Short | Medium | Long |

As can be seen from the coach's test plan, the full factorial experiment would require nine tests, or nine games. Furthermore, although the full factorial experiment would be the effective way to proceed, there are not nine games left because the season is half over. So, although his theory to apply full factorial experimental design is logical, the timing of his introduction of this concept is not optimal. The coach might be better off to wait until preseason to use test his theory.

This hypothetical example also gives an idea as to one of the realities that must occur with not only full factorial experiments but any experiments for that matter: timing and effective planning.

Another reality of any experimental design (or any process improvement initiative for that matter) is that of **noise**. Noise is the effect on the output characteristic that is caused by uncontrollable factors. For example, in the football coach's situation, noise factors include dramatic changes in the weather. Although noise is apparent in any experiment, Dr. Taguchi, as discussed in Section 12.5.4, offers through his design of experiments theory a methodology for addressing noise and the effects on the experiment.

## 12.5.4 Fractional Factorial Experiments

Since the 1940s, a Japanese quality expert named **Dr. Genichi Taguchi** has been involved in quality improvement, introducing several statistical techniques that have been proven to be of extreme value. He emphasized with his techniques the building of quality into the design of products and processes, which is called *Off-Line Quality Control* (whereas *On-Line Quality Control* focuses on the classical aspects such as process control).

Taguchi likewise had a goal of developing a methodology that would decrease the cost and improve the efficiency of experimental design. In order to achieve this he utilized a concept called Fractional Factorials. Fractional Factorials are exactly what they sound like: experiments that require a fraction of the total experiments associated with full factorial experiments. Although the concept of fractional factorials may sound like it compromises integrity when compared to full factorials, Dr. Taguchi's technique focuses on testing only the most meaningful combinations. To do this, Dr. Taguchi built on the Pareto Principle and the Dr. Deming observation that 85% of the bad quality was attributable to the production process while only 15% was attributable to the workers. It was the usage of these theories that provided the nucleus of isolating the main contributing factors to product quality.

When most people think of Dr. Taguchi, they think of his statistical methods, but it is his quality philosophy that needs to be understood and emphasized because it is this philosophy that sets the stage for quality improvement and **robust design**. A robust product design is a design that is relatively insensitive to outside factors. Dr. Taguchi's philosophy focuses around two main thrusts:

1. Quality is measured by the deviation from a specified target value.
2. Quality must be built into the product and process. It can never be "inspected in" or continually reworked to bring it in compliance with specifications. The product must be robust and the process that yields the product must be able to produce that quality product.

In looking at Taguchi's first point, it can be seen that this concept is different from just compliance to specifications. His concept takes quality to the next step, as can be seen in Figure 12–4.

Dr. Taguchi's experimental design theory also addresses noise. By conducting repetitions of the experiment under controlled noise conditions it can be seen that the variation exhibited under these repetitions will be due to the effect of noise. Taguchi expanded the experimental design to include this controlled noise condition, theorizing that optimum conditions (or Taguchi's concept of robust design) that are immune or insensitive to the influence of noise factors can indeed be found. He used what he called an outer array to test the noise conditions, while he tested the various factors and interactions in what he called an inner array.

Although entire textbooks and courses are devoted to the Taguchi methodology and his associated mathematical theories of using orthogonal arrays as the basis for fractional experiments to satisfy a number of situations, it is not the intent of this text to present that level of comprehension. Notwithstanding, Table 12–2 gives an idea of the difference between the number of tests for full and fractional factorial experiments.

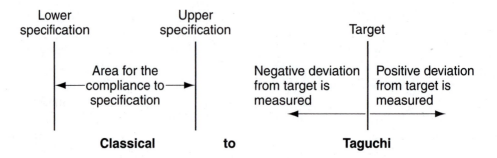

**FIGURE 12–4**
Taguchi's Deviation from Target Concept

**TABLE 12–2**
Comparison of Number of Runs Between Full Factorial and Fractional Factorial

| FACTORS | LEVELS | FULL FACTORIAL | TAGUCHI FRACTIONAL |
|---------|--------|----------------|--------------------|
| 7 | 2 | 128 | 8 |
| 11 | 2 | 2048 | 12 |
| 15 | 2 | 32768 | 16 |
| 31 | 2 | 2147483648 | 32 |
| 4 | 3 | 81 | 9 |
| 13 | 3 | 1594323 | 27 |
| 5 | 4 | 1024 | 16 |
| 6 | 5 | 15625 | 25 |

As can be seen from Table 12–2 fractional factorials greatly abbreviate the number of tests that need to be conducted. Because the most meaningful combinations are tested, minimal integrity is lost. The fractional factorial concept has definitely brought the application of design of experiments to a more realistic level in terms of complexity, expense, and time. To give a further feeling for the great impact of Taguchi methods on investment of resources, consider a very simple two-factor, three-level experiment as shown in the following example.

**EXAMPLE 1**    In a stamping department, costs associated with nonconforming parts are skyrocketing. Stamped laminations are produced on various punch presses in the department. Some of these punch presses use three-punch dies, some use four-punch dies, and some use five-punch dies for stamping out the rotors for small industrial motors. The problem is that the laminations are going out of tolerance very quickly. The foreman and job setter have some empirical data that indicate, along with their experienced hunches, that the problem focuses around the punch dies, the speed of the punch press, and the amount of cooling lubricant. The foreman and job setter feel that the answer lies in conducting a design of experiments. They feel that the factors and levels are those shown as follows:

| FACTOR | LEVEL 1 | LEVEL 2 |
|---|---|---|
| Punch press dies | Three-punch | Four-punch |
| Punch press speed | 40 per minute | 50 per minute |
| Lubricant | 3 quarts/press | 4 quarts/press |

The full factorial design for this experiment would require eight runs as shown below:

| EXPERIMENT | DIES | SPEED | LUBRICANT AMOUNT |
|---|---|---|---|
| 1 | Level 1 | Level 1 | Level 1 |
| 2 | Level 1 | Level 1 | Level 2 |
| 3 | Level 1 | Level 2 | Level 1 |
| 4 | Level 1 | Level 2 | Level 2 |
| 5 | Level 2 | Level 1 | Level 1 |
| 6 | Level 2 | Level 1 | Level 2 |
| 7 | Level 2 | Level 2 | Level 1 |
| 8 | Level 2 | Level 2 | Level 2 |

Replacing the actual settings for the levels of the experiment yields:

| EXPERIMENT | DIES | SPEED | LUBRICANT AMOUNT |
|---|---|---|---|
| 1 | three | 40/min | 3 quarts |
| 2 | three | 40/min | 4 quarts |
| 3 | three | 50/min | 3 quarts |
| 4 | three | 50/min | 4 quarts |
| 5 | four | 40/min | 3 quarts |
| 6 | four | 40/min | 4 quarts |
| 7 | four | 50/min | 3 quarts |
| 8 | four | 50/min | 4 quarts |

However, in looking at the fractional factorial Taguchi test plan layout, it can be seen that only four experiments would need to be conducted:

| EXPERIMENT | DIES | SPEED | LUBRICANT AMOUNT |
|---|---|---|---|
| 1 | Level 1 | Level 1 | Level 1 |
| 2 | Level 1 | Level 2 | Level 2 |
| 3 | Level 2 | Level 1 | Level 2 |
| 4 | Level 2 | Level 2 | Level 1 |

And replacing the levels with the actual settings yields:

| EXPERIMENT | DIES | SPEED | LUBRICANT AMOUNT |
|---|---|---|---|
| 1 | three | 40/min | 3 quarts |
| 2 | three | 50/min | 4 quarts |
| 3 | four | 40/min | 4 quarts |
| 4 | four | 50/min | 3 quarts |

As can be seen by comparing the full factorial design to the Taguchi design the number of experiments is greatly reduced. In reality, most problems involve more than just three factors and two levels, so the power and advantage of the Taguchi method over other types of experiments is very evident.

It must be again restated and cautioned, however, that when considering the Taguchi methodology it is not the intention of this text to go into the complete depth of all of the aspects of the Taguchi theory. There are many more aspects and considerations in a Taguchi design than demonstrated previously. For example, the third column (lubricant amount) also represents the interaction of the first two factors (dies and speed). In addition, the signal-to-noise aspect has not been shown. However, the preceding example does set the stage for the advantages of the Taguchi methodology.

## 12.6  Prerequisite Steps in Conducting an Experiment

There are several prerequisite planning steps involved in conducting an experiment. These steps are vital because even though the benefits of conducting an experiment are great, it is still a reality that conducting an experiment requires the expenditure of organizational resources. Effective planning will not only assist in gaining management support for the conducting of the experiment, it will also greatly reduce the time and resources required. If these prerequisite steps are not followed exhaustively, then there is great risk of conducting a haphazard experiment and organizational resources will have been wasted. Wasted resources, in turn, will definitely put an end to future endeavors involving experimental design (as well as any career advancements in the world of quality engineering).

### Prerequisite 1: Cost Analysis and Selection of a Product or Specific Process

The selection of an area or specific process is the initial question that needs to be answered. A process that needs increased performance is usually an ideal selection, but now the question becomes "Which process needs the most improvement?" Insight into process selection is given by usually looking at the cost of poor quality. Recall from Chapter 1 how the many costs of poor quality can quickly add up. Conducting a cost analysis of the products to determine the most beneficial product or process to conduct an experiment would be wise to determine where the biggest benefits could result.

### Prerequisite 2: Develop an Estimate of the Cost of Conducting the Experiment

Conducting an experiment consumes resources and it is very desirable to establish an estimate of the cost associated with the experiment. Replication of the trial runs and experiments that require destructive testing are costs that can add up very quickly in the total cost of the experiment. It is absolutely imperative to obtain an idea as to just how many resources will be required and what the costs will be before ever commencing with the experiment.

### Prerequisite 3: Gain Management Approval and Support to Conduct the Experiment

Without a doubt, it is vitally essential to gain management approval and support before ever commencing with an experiment. Conducting an experiment without management consent can result in the end of the experiment and one's career because it requires the expenditure of resources. Unless someone has free reign to making decisions and authorizing budget spending, it is extremely wise to ensure that management is totally in agreement and on board with the conducting of an experiment. It is very wise to brief management on the cost analysis (Step 1) and experiment cost (Step 2), as well as the details of the experiment before ever commencing with the experiment.

## 12.7  Steps in Conducting an Experiment

Now that the prerequisites have been satisfied, the experiment is ready to be conducted. An experiment requires specific steps that must be taken in a specific order. The specific steps for conducting an experiment are detailed next and discussed in global terms with a follow-on example to demonstrate their application.

- DETERMINE THE IDEAL FUNCTION. The ideal function needs to be identified in specific terms. To determine the ideal function is essential to determine what specifically we want the product or process to accomplish. Although this seems relatively easy, the specificity requires that careful consideration be given.

- SELECT THE QUALITY CHARACTERISTIC FOR THE IDEAL FUNCTION. Evaluation of the ideal function occurs at this point. This is best answered by determining "how" we will know that the ideal function has been achieved. Similar to the determination of a specific ideal function, the quality characteristic must be specific and measurable.
- DETERMINE SIGNIFICANT FACTOR BY BRAINSTORMING. This step is where the Pareto diagram and the Fishbone diagram are extremely beneficial. These two techniques, along with brainstorming by personnel who are very familiar with the product, are instrumental in determining all significant factors to be considered in the experiment.
- SELECT THE FACTORS AND LEVELS. In this step, the factors for inclusion in the test are selected and the levels for these factors are determined. It is critical to determine the largest contributing factors to be tested and not to include factors that do not contribute significantly. The reason for this is that as has been seen above, the more factors that are included in the experiment, the larger the experiment becomes and the more costly it will be. While it may be argued that the more factors that are included, the more robust the product design will be, it must be realized that more factors result in a more complex and costly experimental design.

In a typical experiment, there are usually two levels for each factor: a high level and a low level. However, there are some experiments where three levels are desirable: a low level, a medium level, and a high level. Three level tests are used when it is determined that there exists a nonlinear relationship (possibly parabolic) between the low and high settings of the factor. A two-factor test, on the other hand, is used when it is determined that a linear relationship exists between the low and high level settings of the factor. The concept of using a three-level experiment versus a two-level experiment is shown in Figure 12–5.

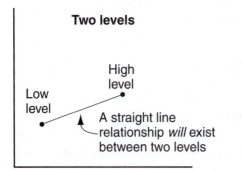 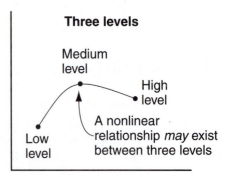

**FIGURE 12–5**
Linearity of Two-Level Experiments Versus Three-Level Experiments

- DEVELOP THE MULTIFACTOR TEST PLAN. This step involves the development of the test plan for conducting the experiment. It details the different combinations of factors and levels for each individual test in the experiment. To do this, **design arrays** are used. A design array is a predetermined array that represents the experimental settings. *The row represents the runs and the columns represent the factors.* Although there are many different design arrays due to the number of factors and levels that can be selected, some of the more common design arrays are shown in Figure 12–6.
- REFINE THE TEST PLAN AND DEVELOP A DATA COLLECTION SHEET. Now that the experiment is taking shape, the test plan is further defined in terms of specificity and a data collection sheet for the experiment is created.

### Three-factor, two-level array factors

|   | 1 | 2 | 3 |
|---|---|---|---|
| 1 | −1 | −1 | −1 |
| 2 | −1 | +1 | +1 |
| 3 | +1 | −1 | +1 |
| 4 | +1 | +1 | −1 |

−1 = Low level
+1 = High level

### Seven-factor, two-level array factors

|   | 1 | 2 | 3 | 4 | 5 | 6 | 7 |
|---|---|---|---|---|---|---|---|
| 1 | −1 | −1 | −1 | −1 | −1 | −1 | −1 |
| 2 | −1 | −1 | −1 | +1 | +1 | +1 | +1 |
| 3 | −1 | +1 | +1 | −1 | −1 | +1 | +1 |
| 4 | −1 | +1 | +1 | +1 | +1 | −1 | −1 |
| 5 | +1 | −1 | +1 | −1 | +1 | −1 | +1 |
| 6 | +1 | −1 | +1 | +1 | −1 | +1 | −1 |
| 7 | +1 | +1 | −1 | −1 | +1 | +1 | −1 |
| 8 | +1 | +1 | −1 | +1 | −1 | −1 | +1 |

### Eleven-factor, two-level array factors

|   | 1 | 2 | 3 | 4 | 5 | 6 | 7 | 8 | 9 | 10 | 11 |
|---|---|---|---|---|---|---|---|---|---|----|----|
| 1 | −1 | −1 | −1 | −1 | −1 | −1 | −1 | −1 | −1 | −1 | −1 |
| 2 | −1 | −1 | −1 | −1 | −1 | +1 | +1 | +1 | +1 | +1 | +1 |
| 3 | −1 | −1 | +1 | +1 | +1 | −1 | −1 | −1 | +1 | +1 | +1 |
| 4 | −1 | +1 | −1 | +1 | +1 | −1 | +1 | +1 | −1 | −1 | +1 |
| 5 | −1 | +1 | +1 | −1 | +1 | +1 | −1 | +1 | −1 | +1 | −1 |
| 6 | −1 | +1 | +1 | +1 | −1 | +1 | +1 | −1 | +1 | −1 | −1 |
| 7 | +1 | −1 | +1 | +1 | −1 | −1 | +1 | +1 | −1 | +1 | −1 |
| 8 | +1 | −1 | +1 | −1 | +1 | +1 | +1 | −1 | −1 | −1 | +1 |
| 9 | +1 | −1 | −1 | +1 | +1 | +1 | −1 | +1 | +1 | −1 | −1 |
| 10 | +1 | +1 | +1 | −1 | −1 | −1 | −1 | +1 | +1 | −1 | +1 |
| 11 | +1 | +1 | −1 | +1 | −1 | +1 | −1 | −1 | −1 | +1 | +1 |
| 12 | +1 | +1 | −1 | −1 | +1 | −1 | +1 | −1 | +1 | +1 | −1 |

**FIGURE 12–6**

*(continued)*

Design Arrays for Different Factor and Level Combinations

### Four-factor, three-level array factors

|   | 1 | 2 | 3 | 4 |
|---|---|---|---|---|
| 1 | 1 | 1 | 1 | 1 |
| 2 | 1 | 2 | 2 | 2 |
| 3 | 1 | 3 | 3 | 3 |
| 4 | 2 | 1 | 2 | 3 |
| 5 | 2 | 2 | 3 | 1 |
| 6 | 2 | 3 | 1 | 2 |
| 7 | 3 | 1 | 3 | 2 |
| 8 | 3 | 2 | 1 | 3 |
| 9 | 3 | 3 | 2 | 1 |

1 = Low level
2 = Medium level
3 = High level

### Eight-factor, three-level array factors

|    | 1 | 2 | 3 | 4 | 5 | 6 | 7 | 8 |
|----|---|---|---|---|---|---|---|---|
| 1  | 1 | 1 | 1 | 1 | 1 | 1 | 1 | 1 |
| 2  | 1 | 1 | 2 | 2 | 2 | 2 | 2 | 2 |
| 3  | 1 | 1 | 3 | 3 | 3 | 3 | 3 | 3 |
| 4  | 1 | 2 | 1 | 1 | 2 | 2 | 3 | 3 |
| 5  | 1 | 2 | 2 | 2 | 3 | 3 | 1 | 1 |
| 6  | 1 | 2 | 3 | 3 | 1 | 1 | 2 | 2 |
| 7  | 1 | 3 | 1 | 2 | 1 | 3 | 2 | 3 |
| 8  | 1 | 3 | 2 | 3 | 2 | 1 | 3 | 1 |
| 9  | 1 | 3 | 3 | 1 | 3 | 2 | 1 | 2 |
| 10 | 2 | 1 | 1 | 3 | 3 | 2 | 2 | 1 |
| 11 | 2 | 1 | 2 | 1 | 1 | 3 | 3 | 2 |
| 12 | 2 | 1 | 3 | 2 | 2 | 1 | 1 | 3 |
| 13 | 2 | 2 | 1 | 2 | 3 | 1 | 3 | 2 |
| 14 | 2 | 2 | 2 | 3 | 1 | 2 | 1 | 3 |
| 15 | 2 | 2 | 3 | 1 | 2 | 3 | 2 | 1 |
| 16 | 2 | 3 | 1 | 3 | 2 | 3 | 1 | 2 |
| 17 | 2 | 3 | 2 | 1 | 3 | 1 | 2 | 3 |
| 18 | 2 | 3 | 3 | 2 | 1 | 2 | 3 | 1 |

**FIGURE 12–6** (*cont'd.*)

- CONDUCT THE EXPERIMENT. The experiment is conducted and the data collection sheet completed for the results of the experimental runs.
- DETERMINE THE OPTIMUM LEVELS AND MAXIMUM EXPECTED RESULTS. After the experiment has been conducted, the optimum levels need to be computed and the maximum expected results at these settings is determined.

## 12.8  Design of Experiments Example

Tomato production has always been of utmost concern to the weekend gardener. In fact, the desire to produce the largest tomatoes of any of the other neighboring gardeners is a glamorous claim to fame. A local gardener decides to conduct an experiment to determine the optimum conditions that will result in the largest tomatoes, thus giving him the bragging rights for the neighborhood. With this in mind, the gardener decides to conduct an experiment, following the steps described previously.

- DETERMINE THE IDEAL FUNCTION. The ideal function for this simple experiment is to grow the largest tomatoes. More specifically, the gardener has a preference of Better Boy tomatoes, so the type of tomato is now more narrowly defined.
- SELECT THE QUALITY CHARACTERISTIC FOR THE IDEAL FUNCTION. The term "largest tomato" of the ideal function needs to be further defined. Does the term mean the biggest in physical dimensions? Or does the term mean the heaviest tomato in terms of weight? The gardener decides that the quality characteristic will be the average circumference of all of the tomatoes at their widest point (because a tomato is not perfectly circular).
- DETERMINE SIGNIFICANT FACTOR BY BRAINSTORMING. Over the years, the gardener has compiled an extremely large wealth of information about growing tomatoes and feels that he can determine the significant factors that contribute to tomato growth. Notwithstanding, he feels that to make the most of this experiment he should develop the list along with a couple of his friends at the local nursery who are extremely knowledgeable and have vast experience with tomatoes as well. The brainstormed significant factors include the following:

  - The amount of phosophorous in the fertilizer
  - The number of times the plant is fertilized in a growing season
  - The amount of daily water
  - The time of daily watering (morning, midday, or evening)
  - The amount of pruning (no pruning of suckers to heavy pruning of suckers. A sucker is a stem that produces no blooms, but does absorb energy from the rest of the plant.)
  - Whether the plants are staked, semistaked, or left to grow on the ground
  - The hours of direct light
  - The quality of the Better Boy plants
  - The amount of insecticide used to control insects

- SELECT THE FACTORS AND LEVELS. The gardener decided that due to the size of his garden and the amount of effort he wanted to spend on his experiment, that he would limit his experiment to three factors with two levels for each factor. The gardener and his nursery friends narrowed the factors to the three main factors and determined the levels for each factor as follows:

| CONTROL FACTORS | LOW LEVEL (−1) | HIGH LEVEL(+1) |
| --- | --- | --- |
| 1. Daily water amount | 1 cup | 3 cups |
| 2. Pruning | None | Total |
| 3. Phosphorous | 20% | 40% |

- DEVELOP THE MULTIFACTOR TEST PLAN. The test plan that will be used is an L4 Array because there are three factors requiring four runs. The factors are assigned to the columns, with a −1 representing the low-level setting and a +1 representing the high-level setting.

| RUN # | WATER | PRUNING | PHOSPHOROUS |
| --- | --- | --- | --- |
| 1 | −1 | −1 | −1 |
| 2 | −1 | 1 | 1 |
| 3 | 1 | −1 | 1 |
| 4 | 1 | 1 | −1 |

Note in the multifactor test plan that all of the factors are tested equally. There are two runs where the water is tested at the low level and two runs where water is tested at the high level. There are two runs where the pruning is tested at the low level and two runs where pruning is tested at the high level. And finally, there are two runs where the phosphorous is tested at the low level and two runs where phosphorous is tested at the high level. This is referred to as a balanced design, which is one of the key concepts of two-level designs.

- REFINE THE TEST PLAN AND DEVELOP A DATA COLLECTION SHEET. The experiment is just about ready to be conducted. The gardener refines the test plan by substituting the actual wording for the levels and develops a data collection sheet to record the results.

| RUN # | WATER | PRUNING | PHOSPHOROUS | RESULT |
| --- | --- | --- | --- | --- |
| 1 | 1 cup | None | 20% | |
| 2 | 1 cup | Total | 40% | |
| 3 | 3 cups | None | 40% | |
| 4 | 3 cups | Total | 20% | |

- CONDUCT THE EXPERIMENT. The experiment is now conducted during the growing season and the results (average circumference of all tomatoes) are calculated and shown in the results column.

| RUN # | WATER | PRUNING | PHOSPHOROUS | RESULT |
|-------|-------|---------|-------------|--------|
| 1 | 1 cup | None | 20% | 5.6 in. |
| 2 | 1 cup | Total | 40% | 10.9 in. |
| 3 | 3 cups | None | 40% | 9.7 in. |
| 4 | 3 cups | Total | 20% | 8.2 in. |

It can be seen from conducting the experiment that run 2 resulted in the greatest average circumference. However, it must be remembered that there were only four runs conducted (of which 2 was the best) and that not all of the different combinations were tested. The next step explores the determination of the optimum level based on results obtained from conducting the experiment.

- DETERMINE THE OPTIMUM LEVELS AND MAXIMUM EXPECTED RESULTS. The results from the experiment show that the second run yielded the largest average circumference (10.9 in.) for the factors and levels that were tested. But the experiment does not stop there. The optimum levels (which may not necessarily be the ones tested in the experiment) need to be determined and the expected results from the optimum level settings determined. To do this, several steps are required. First, the multilevel test plan is revisited and the results incorporated as well.

| RUN # | WATER | PRUNING | PHOSPHOROUS | RESULT |
|-------|-------|---------|-------------|--------|
| 1 | −1 | −1 | −1 | 5.6 in. |
| 2 | −1 | 1 | 1 | 10.9 in. |
| 3 | 1 | −1 | 1 | 9.7 in. |
| 4 | 1 | 1 | −1 | 8.2 in. |

The equation for the prediction equation is:

$$\text{OUTPUT} = X1 + X2F1 + X3F2 + X4F3$$

where: $X_1$ = the average of the four run results
$X_2$ = the coefficient for the first factor (water)
$X_3$ = the coefficient for the second factor (pruning)
$X_4$ = the coefficient for the third factor (phosphorous)
$F_1$ = the first factor
$F_2$ = the second factor
$F_3$ = the third factor

To complete the prediction equation for the experiment, the variables are now computed. The first variable (X1) is the average of the four results:

$$X_1 = \frac{(5.6 + 10.9 + 9.7 + 8.2)}{4} = 8.6 \text{ in.}$$

To compute the coefficient for the first factor (water), the column settings (1, −1) for water in each run are multiplied by the result of each run, and then an average is obtained:

$$X_2 = \frac{\left[(-1)(5.6) + (-1)(10.9) + (1)(9.7) + (1)(8.2)\right]}{4} = 0.35$$

The coefficients for each of the remaining factors is computed the same way as the first factor:

$$X_3 = \frac{\left[(-1)(5.6) + (1)(10.9) + (-1)(9.7) + (1)(8.2)\right]}{4} = 0.95$$

$$X_4 = \frac{\left[(-1)(5.6) + (1)(10.9) + (1)(9.7) + (-1)(8.2)\right]}{4} = 1.7$$

The equation for the experiment can now be completed:

$$\text{OUTPUT} = 8.6 + 0.35F_1 + 0.95F_2 + 1.7F_3$$

This equation is the equation for the experiment that was conducted. To prove this, one of the runs is selected and the values tested in the equation. Run 3 yielded an output of 9.7 in. The setting for Factor 1 in this run was +1. The setting for Factor 2 in this run was −1. And the setting for Factor 3 in this run was +1. Substituting these values into the equation yields:

$$\text{OUTPUT} = 8.6 + (0.35)(1) + (0.95)(-1) + (1.7)(1)$$

$$\text{OUTPUT} = 9.7 \text{ in.}$$

Note that the answer obtained confirms the results obtained in the experiment for run 3. The experiment prediction equation will work for any of the runs in the experiment.

Now that the equation for the experiment has been derived, the determination of the optimum levels (including levels that were not tested in the experiment) can be conducted. The experiment was conducted and the output equation determined for the factors was tested. Using the results of the experiment, it is time to determine settings of the factors that will yield the greatest yield. But the optimum settings may be ones that were not tested in the experiment. This is one of the great benefits of fractional factorial experiments. Via all of the hidden mathematics associated with experiments it is now possible to determine the theoretically best combination and then test the results. And restating, although the in depth mathematics will not be covered in this text, the mathematics associated with design of experiments does support the approach to determining the optimum level settings for yielding the greatest output.

To obtain the optimum settings, the experiment equation needs to be revisited:

$$OUTPUT = 8.6 + 0.35F_1 + 0.95F_2 + 1.7F_3$$

In studying the experiment equation it can be seen that in order to obtain the maximum output, all of the terms need to be positive. In other words, the first term (8.6) is positive, and we want to add the remaining terms, not subtract them. Term 2 ($+0.35F1$) will be negative if the setting of the first factor is a $-1$. If this happens, then the output will be decreased (the average tomato circumference will be 0.35 in. less in circumference). Therefore, the setting selected for factor 1 will be $+1$. The setting for factor 2 will be $+1$ to obtain a positive second term, and the setting for factor 3 will be $+1$. The optimum level settings and maximum expected results now become:

$$MAXIMUM\ OUTPUT = 8.6 + 0.35F_1 + 0.95F_2 + 1.7F_3$$
$$MAXIMUM\ OUTPUT = 8.6 + (0.35)(+1) + (0.95)(+1) + (1.7)(+1)$$
$$MAXIMUM\ OUTPUT = 11.6\ in.$$

Two things can be observed from determining the optimum level settings for maximum expected output. The first is that the maximum expected output is 11.6 inches as opposed to the 10.9 in. from the experiment. The second is that the settings for the maximum expected yield ($+1, +1, +1$) were never tested in the original experiment. Naturally, a confirmation run would be conducted to validate the experimental results. But due to the concept of experimental design and the associated mathematics, the confirmation run with the new levels would be relatively close to the expected result of 11.6 in.

The preceding is a simple example to demonstrate the concept of experimental design. There is much more to design of experiments, such as noise outer arrays and interaction of factors, that are included in the total concept of experiments, but armed with an appreciation of the design of experiments concept, the student can appreciate this technique as a tool in the process improvement toolbag.

# 12.9 Summary

- Design of Experiments is the manipulation of controllable factors in a systematic fashion to not only determine their effect on some desired outcome, but to find the optimum combination of factors that will yield the maximum desired output.
- There are four types of experiments covered in this text: trial and error, one factor at a time, full factorial, and fractional factorial.
- Trial and error experimentation is where one factor is manipulated without regard to the other factors.
- One at a time experimentation is where a series of factors are studied, but only one factor is adjusted at a time while the others are held constant.
- Full factorial experimentation involves considering all factors and all combinations of those factors.
- The formula for determining the number of tests required in a full factorial experiment is:

$$L^f$$

where: L = the number of levels for the factors

f = the number of factors in the experiment

- Fractional factorial experimentation involves considering only the largest contributing factors in the experiment.
- A three-level experiment is selected over a two-level experiment when it is important to know if there is a linear relationship between the low- and high-level settings of the factors.
- It is vital to work through all of the prerequisite steps before conducting an experiment. Selection of the product, cost analysis of conducting the experiment, and gaining management support and approval for the experiment are included in the prerequisite steps.
- It is equally vital to conduct the experiment in an orderly manner as depicted in the steps for conducting an experiment.
- The maximum output and level settings for an experiment may include levels that were not tested in the fractional factorial experiment.

## Review Questions

1. A full factorial experiment requires test runs for:
   a. some of the factors at all of their levels
   b. some of the factors at some of their levels

   c. all of the factors at all of their levels
   d. all of the factors at some of their levels
   e. none of the above

2. Dr. Taguchi is associated with the theory of:
   a. fractional factorials
   b. full factorials
   c. balanced designs
   d. none of the above

3. An eleven-factor, two-level full factorial experiment would require how many runs?
   a. 128
   b. 16
   c. 81
   d. None of the above

4. The prerequisite steps for conducting an experiment include:
   a. selecting factors and levels
   a. determining the ideal function
   b. developing an estimate of the cost of conducting the experiment
   c. all of the above
   d. none of the above

5. A fractional factorial experiment requires test runs for:
   a. some of the factors at all of their levels
   b. some of the factors at some of their levels
   c. all of the factors at all of their levels
   d. all of the factors at some of their levels
   e. none of the above

6. Which of the following are included in the steps for conducting an experiment?
   a. Gain management approval and support to conduct the experiment.
   b. Determine significant factor by brainstorming.
   c. Determine the optimum levels and maximum expected results.
   d. Both b and c.
   e. None of the above.

7. The question "How will we know that the ideal function has been achieved?" is part of which of the steps in conducting an experiment?
   a. Develop the multifactor test plan.
   b. Refine the test plan and develop a data collection sheet.
   c. Select the quality characteristic for the ideal function.
   d. None of the above.

8. Design arrays are:
   a. predetermined arrays that represent the experimental factor settings
   b. an instrumental tool for selecting factors
   c. part of the prerequisite steps for conducting an experiment
   d. none of the above

9. Design of experiments:
   a. is always associated with full factorial experimentation
   b. was invented by Dr. Taguchi
   c. is concerned with manipulating controllable factors in order to determine their effect on some desired outcome
   d. none of the above

10. The type of experimentation where one factor is manipulated without regard to other factors is:
    a. full factorial
    b. trial and error
    c. one factor at a time
    d. none of the above

11. The type of experimentation that may study a series of factors but only adjusts one factor at a time is called:
    a. full factorial
    b. fractional factorial
    c. trial and error
    d. none of the above

12. One aspect of Dr. Taguchi's philosophy regarding quality includes:
    a. quality must be built into the product and process
    b. quality is measured by the deviation from a specified target value
    c. both a and b
    d. none of the above

13. If an experiment has five factors and three levels for each factor, how many runs would be needed to conduct a full factorial experiment?
    a. 125
    b. 243
    c. 625
    d. None of the above

14. If an experiment has twelve factors and two levels for each factor, how many runs would be needed to conduct a full factorial experiment?
    a. 4096
    b. 144

   c. 3375
   d. None of the above

15. When conducting an experiment, it is always best to include as many factors as possible. True or False

16. One of the biggest disadvantages to conducting full factorial experiments is:
   a. the mathematics of full factorial experiments is very difficult and time consuming
   b. the high costs of experiments that have many factors
   c. there are no disadvantages—full factorial experimentation is very thorough
   d. none of the above

17. The disadvantages associated with trial and error are:
   a. it is slow
   b. it is inefficient
   c. it is costly
   d. all of the above
   e. none of the above

18. A fifteen-factor, two-level Taguchi fractional factorial experiment would require how many runs?
   a. 1206
   b. 1600
   c. 3208
   d. None of the above

19. A thirteen-factor, three-level Taguchi fractional factorial experiment would require how many runs?
   a. 27
   b. 25
   c. 16
   d. None of the above

20. Three level experiments are advantageous over two-level experiments whenever:
   a. time and money exist to support the experiment
   b. the difference between the low and high levels is so great that a middle level would be beneficial
   c. it is important to determine if a nonlinear relationship exists between the low and high levels
   d. none of the above

## Problems

1. Consider making a balsa wood airplane and flying it in competition. The airplane, which will be no longer than 2 ft. in length, will be entered in a local competition where planes are flown for distance and the top three designs receive awards. State the ideal function. Questions 2 through 9 also apply to this balsa wood airplane experiment.

2. In your opinion, what should the quality characteristic for the ideal function be?

3. In your opinion, what are the significant factors? State your reasoning for each factor.

4. a. Select the significant factors you will choose for the experiment and explain your reasoning.
   b. Select the levels for each factor.
   c. Why did you choose two-level or three-level settings for the experiment?

5. Develop the multifactor test plan.

6. Refine the test plan and develop a data collection sheet for conducting the experiment.

7. Use random numbers (between 1 and 50) that are generated using the software in the back of the text to represent the results of the experiment for each test. Note: the random numbers represent the distance flown in feet.

8. Determine the output equation and coefficients for each factor. Show all of your work.

9. a. Determine the maximum output equation and optimum level settings for the equation.
   b. Are the level settings for the maximum output the same as any that were tested?

10. Think of an application where you would apply a design of experiments.
    a. State the ideal function.
    b. Select the quality characteristic and explain your reasoning.
    c. Brainstorm the significant factors.
    d. Select the factors for the experiment and explain your reasoning.
    e. Select the level settings for each factor.
    f. Develop a multifactor test plan for the experiment.

# Appendix

**TABLE 1**

500 Uniformly Distributed Five-Digit Random Numbers

| | | | | | | | | | |
|---|---|---|---|---|---|---|---|---|---|
| 29450 | 69447 | 35691 | 88555 | 94265 | 76964 | 01218 | 45909 | 05905 | 54405 |
| 41349 | 79445 | 29655 | 51698 | 19046 | 12030 | 11594 | 36497 | 63618 | 08939 |
| 46397 | 03342 | 05765 | 85121 | 17975 | 64219 | 68349 | 45503 | 41084 | 36253 |
| 65883 | 98919 | 82033 | 92034 | 15992 | 36301 | 86641 | 88298 | 15879 | 13739 |
| 93517 | 38563 | 86977 | 31068 | 60838 | 45301 | 40858 | 46559 | 84719 | 57325 |
| | | | | | | | | | |
| 69603 | 18406 | 37309 | 68004 | 13666 | 13568 | 02741 | 19904 | 10437 | 74632 |
| 39063 | 31516 | 76564 | 85289 | 44257 | 11667 | 27207 | 25095 | 54506 | 77266 |
| 85335 | 54109 | 52784 | 39970 | 84175 | 18009 | 15952 | 77458 | 15952 | 14389 |
| 78435 | 54856 | 30045 | 10776 | 40632 | 12021 | 33698 | 55043 | 00192 | 82253 |
| 98351 | 68828 | 34229 | 58067 | 24817 | 64595 | 10681 | 69691 | 70335 | 76936 |
| | | | | | | | | | |
| 56871 | 91841 | 90663 | 48768 | 63423 | 44492 | 00357 | 63045 | 51664 | 03171 |
| 04318 | 60573 | 82610 | 84484 | 03766 | 34614 | 35694 | 22431 | 29749 | 42866 |
| 35355 | 01193 | 07309 | 27430 | 53559 | 59977 | 03449 | 83446 | 66481 | 35554 |
| 62007 | 66942 | 93367 | 58125 | 78569 | 82613 | 95772 | 66115 | 18552 | 42079 |
| 65263 | 55574 | 31565 | 33271 | 77736 | 29731 | 65199 | 91603 | 42439 | 64390 |
| | | | | | | | | | |
| 31339 | 13486 | 23877 | 54777 | 28156 | 02902 | 64659 | 59437 | 70875 | 91570 |
| 87465 | 06845 | 34467 | 23359 | 04254 | 80336 | 61381 | 43653 | 10788 | 46955 |
| 37418 | 84142 | 94667 | 45921 | 88933 | 83785 | 34931 | 02243 | 04407 | 36549 |
| 21439 | 81191 | 33610 | 06482 | 72038 | 49183 | 67363 | 00839 | 75752 | 56764 |
| 51145 | 41822 | 40064 | 09271 | 35596 | 27567 | 70415 | 40824 | 17051 | 23029 |

*(continued)*

**TABLE 1** *(cont'd.)*
500 Uniformly Distributed Five-Digit Random Numbers

| | | | | | | | | | |
|---|---|---|---|---|---|---|---|---|---|
| 60063 | 57249 | 71236 | 71382 | 18412 | 63588 | 33274 | 66530 | 15943 | 41792 |
| 27082 | 77388 | 73891 | 94240 | 97329 | 64552 | 76732 | 18708 | 47019 | 79958 |
| 35691 | 30399 | 33930 | 90194 | 81331 | 96666 | 28895 | 14054 | 13358 | 00067 |
| 89162 | 15000 | 85808 | 95366 | 49916 | 39082 | 55665 | 76085 | 87081 | 39048 |
| 73140 | 16266 | 54536 | 75182 | 41612 | 03073 | 48881 | 33521 | 25483 | 25785 |
| | | | | | | | | | |
| 04852 | 52092 | 33955 | 36912 | 74535 | 95781 | 58949 | 50984 | 93489 | 70451 |
| 95223 | 96660 | 86916 | 16953 | 20856 | 87792 | 18985 | 75642 | 66218 | 29035 |
| 38459 | 62013 | 49507 | 21695 | 23707 | 44114 | 07053 | 85356 | 80293 | 62340 |
| 96444 | 60963 | 94204 | 82210 | 00647 | 10993 | 93874 | 20816 | 11222 | 82311 |
| 24125 | 37498 | 18756 | 47328 | 85143 | 29404 | 88674 | 89138 | 39610 | 22593 |
| | | | | | | | | | |
| 50483 | 46202 | 41148 | 99215 | 78670 | 82222 | 87157 | 51493 | 93282 | 96871 |
| 49198 | 03082 | 58387 | 42915 | 84041 | 88405 | 06494 | 65462 | 78502 | 30662 |
| 67085 | 28852 | 31187 | 09067 | 27353 | 46916 | 24296 | 55992 | 03345 | 82149 |
| 88564 | 74586 | 49501 | 29456 | 14502 | 17572 | 84868 | 50117 | 13077 | 17755 |
| 63997 | 71529 | 78868 | 10950 | 70717 | 18448 | 58961 | 41892 | 62852 | 91259 |
| | | | | | | | | | |
| 80586 | 93798 | 46086 | 56944 | 40669 | 32160 | 05020 | 68404 | 10452 | 03473 |
| 69585 | 77687 | 58256 | 79402 | 44938 | 58812 | 43516 | 87133 | 82970 | 05771 |
| 85280 | 16458 | 23520 | 21369 | 58192 | 44270 | 83449 | 62489 | 19184 | 67012 |
| 21689 | 57451 | 17923 | 81825 | 58412 | 61043 | 94201 | 06897 | 76387 | 62886 |
| 15320 | 81743 | 82726 | 14740 | 56718 | 37943 | 59083 | 92394 | 78792 | 81062 |
| | | | | | | | | | |
| 78471 | 57710 | 03198 | 74290 | 16797 | 36549 | 41688 | 64491 | 37238 | 78087 |
| 80553 | 41230 | 75618 | 11234 | 73946 | 69481 | 04868 | 63731 | 49031 | 86525 |
| 06714 | 59587 | 76228 | 03806 | 28730 | 60426 | 07611 | 04236 | 54850 | 66456 |
| 36567 | 75655 | 67488 | 42576 | 81908 | 71059 | 29398 | 85356 | 49494 | 36185 |
| 63359 | 96419 | 74541 | 95141 | 57863 | 77922 | 97692 | 34617 | 76885 | 15720 |
| | | | | | | | | | |
| 78035 | 77416 | 01886 | 65855 | 96129 | 85808 | 84191 | 09979 | 50215 | 85439 |
| 35780 | 95382 | 98705 | 50331 | 36765 | 25144 | 52088 | 39695 | 87520 | 14203 |
| 67381 | 27054 | 83711 | 76787 | 48237 | 12451 | 42274 | 60957 | 03839 | 58220 |
| 56514 | 95839 | 69719 | 39063 | 08414 | 67161 | 43601 | 47007 | 16312 | 86769 |
| 39436 | 41822 | 46040 | 71669 | 12683 | 87636 | 63667 | 98305 | 60054 | 66261 |

**TABLE 2**

Areas Under the Normal Curve

Proportion of total area of the standard normal curve

| z | 0.09 | 0.08 | 0.07 | 0.06 | 0.05 | 0.04 | 0.03 | 0.02 | 0.01 | 0.00 |
|---|------|------|------|------|------|------|------|------|------|------|
| −3.5 | 0.00017 | 0.00017 | 0.00018 | 0.00019 | 0.00019 | 0.00020 | 0.00021 | 0.00022 | 0.00022 | 0.00023 |
| −3.4 | 0.00024 | 0.00025 | 0.00026 | 0.00027 | 0.00028 | 0.00029 | 0.00030 | 0.00031 | 0.00032 | 0.00034 |
| −3.3 | 0.00035 | 0.00036 | 0.00038 | 0.00039 | 0.00040 | 0.00042 | 0.00043 | 0.00045 | 0.00047 | 0.00048 |
| −3.2 | 0.00050 | 0.00052 | 0.00054 | 0.00056 | 0.00058 | 0.00060 | 0.00062 | 0.00064 | 0.00066 | 0.00069 |
| −3.1 | 0.00071 | 0.00074 | 0.00076 | 0.00079 | 0.00082 | 0.00084 | 0.00087 | 0.00090 | 0.00094 | 0.00097 |
| −3.0 | 0.00100 | 0.00104 | 0.00107 | 0.00111 | 0.00114 | 0.00118 | 0.00122 | 0.00126 | 0.00131 | 0.00135 |
| −2.9 | 0.00139 | 0.00144 | 0.00149 | 0.00154 | 0.00159 | 0.00164 | 0.00169 | 0.00175 | 0.00181 | 0.00187 |
| −2.8 | 0.00193 | 0.00199 | 0.00205 | 0.00212 | 0.00219 | 0.00226 | 0.00233 | 0.00240 | 0.00248 | 0.00256 |
| −2.7 | 0.00264 | 0.00272 | 0.00280 | 0.00289 | 0.00298 | 0.00307 | 0.00317 | 0.00326 | 0.00336 | 0.00347 |
| −2.6 | 0.00357 | 0.00368 | 0.00379 | 0.00391 | 0.00402 | 0.00415 | 0.00427 | 0.00440 | 0.00453 | 0.00466 |
| −2.5 | 0.00480 | 0.00494 | 0.00508 | 0.00523 | 0.00539 | 0.00554 | 0.00570 | 0.00587 | 0.00604 | 0.00621 |
| −2.4 | 0.00639 | 0.00657 | 0.00676 | 0.00695 | 0.00714 | 0.00734 | 0.00755 | 0.00776 | 0.00798 | 0.00820 |
| −2.3 | 0.00842 | 0.00866 | 0.00889 | 0.00914 | 0.00939 | 0.00964 | 0.00990 | 0.01017 | 0.01044 | 0.01072 |
| −2.2 | 0.01101 | 0.01130 | 0.01160 | 0.01191 | 0.01222 | 0.01255 | 0.01287 | 0.01321 | 0.01355 | 0.01390 |
| −2.1 | 0.01426 | 0.01463 | 0.01500 | 0.01539 | 0.01578 | 0.01618 | 0.01659 | 0.01700 | 0.01743 | 0.01786 |
| −2.0 | 0.01831 | 0.01876 | 0.01923 | 0.01970 | 0.02018 | 0.02068 | 0.02118 | 0.02169 | 0.02222 | 0.02275 |
| −1.9 | 0.02330 | 0.02385 | 0.02442 | 0.02500 | 0.02559 | 0.02619 | 0.02680 | 0.02743 | 0.02807 | 0.02872 |
| −1.8 | 0.02938 | 0.03005 | 0.03074 | 0.03144 | 0.03216 | 0.03288 | 0.03362 | 0.03438 | 0.03515 | 0.03593 |
| −1.7 | 0.03673 | 0.03754 | 0.03836 | 0.03920 | 0.04006 | 0.04093 | 0.04182 | 0.04272 | 0.04363 | 0.04457 |
| −1.6 | 0.04551 | 0.04648 | 0.04746 | 0.04846 | 0.04947 | 0.05050 | 0.05155 | 0.05262 | 0.05370 | 0.05480 |
| −1.5 | 0.05592 | 0.05705 | 0.05821 | 0.05938 | 0.06057 | 0.06178 | 0.06301 | 0.06426 | 0.06552 | 0.06681 |
| −1.4 | 0.06811 | 0.06944 | 0.07078 | 0.07215 | 0.07353 | 0.07493 | 0.07636 | 0.07780 | 0.07927 | 0.08076 |
| −1.3 | 0.08226 | 0.08379 | 0.08534 | 0.08692 | 0.08851 | 0.09012 | 0.09176 | 0.09342 | 0.09510 | 0.09680 |
| −1.2 | 0.09853 | 0.10027 | 0.10204 | 0.10383 | 0.10565 | 0.10749 | 0.10935 | 0.11123 | 0.11314 | 0.11507 |
| −1.1 | 0.11702 | 0.11900 | 0.12100 | 0.12302 | 0.12507 | 0.12714 | 0.12924 | 0.13136 | 0.13350 | 0.13567 |
| −1.0 | 0.13786 | 0.14007 | 0.14231 | 0.14457 | 0.14686 | 0.14917 | 0.15151 | 0.15386 | 0.15625 | 0.15866 |
| −0.9 | 0.16109 | 0.16354 | 0.16602 | 0.16853 | 0.17106 | 0.17361 | 0.17619 | 0.17879 | 0.18141 | 0.18406 |
| −0.8 | 0.18673 | 0.18943 | 0.19215 | 0.19489 | 0.19766 | 0.20045 | 0.20327 | 0.20611 | 0.20897 | 0.21186 |
| −0.7 | 0.21476 | 0.21770 | 0.22065 | 0.22363 | 0.22663 | 0.22965 | 0.23270 | 0.23576 | 0.23885 | 0.24196 |
| −0.6 | 0.24510 | 0.24825 | 0.25143 | 0.25463 | 0.25785 | 0.26109 | 0.26435 | 0.26763 | 0.27093 | 0.27425 |
| −0.5 | 0.27760 | 0.28096 | 0.28434 | 0.28774 | 0.29116 | 0.29460 | 0.29806 | 0.30153 | 0.30503 | 0.30854 |
| −0.4 | 0.31207 | 0.31561 | 0.31918 | 0.32276 | 0.32636 | 0.32997 | 0.33360 | 0.33724 | 0.34090 | 0.34458 |
| −0.3 | 0.34827 | 0.35197 | 0.35569 | 0.35942 | 0.36317 | 0.36693 | 0.37070 | 0.37448 | 0.37828 | 0.38209 |
| −0.2 | 0.38591 | 0.38974 | 0.39358 | 0.39743 | 0.40129 | 0.40517 | 0.40905 | 0.41294 | 0.41683 | 0.42074 |
| −0.1 | 0.42465 | 0.42858 | 0.43251 | 0.43644 | 0.44038 | 0.44433 | 0.44828 | 0.45224 | 0.45620 | 0.46017 |
| −0.0 | 0.46414 | 0.46812 | 0.47210 | 0.47608 | 0.48006 | 0.48405 | 0.48803 | 0.49202 | 0.49601 | 0.50000 |

*(continued)*

**TABLE 2** *(cont'd.)*

Areas Under the Normal Curve

| z | 0.00 | 0.01 | 0.02 | 0.03 | 0.04 | 0.05 | 0.06 | 0.07 | 0.08 | 0.09 |
|---|------|------|------|------|------|------|------|------|------|------|
| **0.0** | 0.50000 | 0.50399 | 0.50798 | 0.51197 | 0.51595 | 0.51994 | 0.52392 | 0.52790 | 0.53188 | 0.53586 |
| **0.1** | 0.53983 | 0.54380 | 0.54776 | 0.55172 | 0.55567 | 0.55962 | 0.56356 | 0.56749 | 0.57142 | 0.57535 |
| **0.2** | 0.57926 | 0.58317 | 0.58706 | 0.59095 | 0.59483 | 0.59871 | 0.60257 | 0.60642 | 0.61026 | 0.61409 |
| **0.3** | 0.61791 | 0.62172 | 0.62552 | 0.62930 | 0.63307 | 0.63683 | 0.64058 | 0.64431 | 0.64803 | 0.65173 |
| **0.4** | 0.65542 | 0.65910 | 0.66276 | 0.66640 | 0.67003 | 0.67364 | 0.67724 | 0.68082 | 0.68439 | 0.68793 |
| **0.5** | 0.69146 | 0.69497 | 0.69847 | 0.70194 | 0.70540 | 0.70884 | 0.71226 | 0.71566 | 0.71904 | 0.72240 |
| **0.6** | 0.72575 | 0.72907 | 0.73237 | 0.73565 | 0.73891 | 0.74215 | 0.74537 | 0.74857 | 0.75175 | 0.75490 |
| **0.7** | 0.75804 | 0.76115 | 0.76424 | 0.76730 | 0.77035 | 0.77337 | 0.77637 | 0.77935 | 0.78230 | 0.78524 |
| **0.8** | 0.78814 | 0.79103 | 0.79389 | 0.79673 | 0.79955 | 0.80234 | 0.80511 | 0.80785 | 0.81057 | 0.81327 |
| **0.9** | 0.81594 | 0.81859 | 0.82121 | 0.82381 | 0.82639 | 0.82894 | 0.83147 | 0.83398 | 0.83646 | 0.83891 |
| **1.0** | 0.84134 | 0.84375 | 0.84614 | 0.84849 | 0.85083 | 0.85314 | 0.85543 | 0.85769 | 0.85993 | 0.86214 |
| **1.1** | 0.86433 | 0.86650 | 0.86864 | 0.87076 | 0.87286 | 0.87493 | 0.87698 | 0.87900 | 0.88100 | 0.88298 |
| **1.2** | 0.88493 | 0.88686 | 0.88877 | 0.89065 | 0.89251 | 0.89435 | 0.89617 | 0.89796 | 0.89973 | 0.90147 |
| **1.3** | 0.90320 | 0.90490 | 0.90658 | 0.90824 | 0.90988 | 0.91149 | 0.91308 | 0.91466 | 0.91621 | 0.91774 |
| **1.4** | 0.91924 | 0.92073 | 0.92220 | 0.92364 | 0.92507 | 0.92647 | 0.92785 | 0.92922 | 0.93056 | 0.93189 |
| **1.5** | 0.93319 | 0.93448 | 0.93574 | 0.93699 | 0.93822 | 0.93943 | 0.94062 | 0.94179 | 0.94295 | 0.94408 |
| **1.6** | 0.94520 | 0.94630 | 0.94738 | 0.94845 | 0.94950 | 0.95053 | 0.95154 | 0.95254 | 0.95352 | 0.95449 |
| **1.7** | 0.95543 | 0.95637 | 0.95728 | 0.95818 | 0.95907 | 0.95994 | 0.96080 | 0.96164 | 0.96246 | 0.96327 |
| **1.8** | 0.96407 | 0.96485 | 0.96562 | 0.96638 | 0.96712 | 0.96784 | 0.96856 | 0.96926 | 0.96995 | 0.97062 |
| **1.9** | 0.97128 | 0.97193 | 0.97257 | 0.97320 | 0.97381 | 0.97441 | 0.97500 | 0.97558 | 0.97615 | 0.97670 |
| **2.0** | 0.97725 | 0.97778 | 0.97831 | 0.97882 | 0.97932 | 0.97982 | 0.98030 | 0.98077 | 0.98124 | 0.98169 |
| **2.1** | 0.98214 | 0.98257 | 0.98300 | 0.98341 | 0.98382 | 0.98422 | 0.98461 | 0.98500 | 0.98537 | 0.98574 |
| **2.2** | 0.98610 | 0.98645 | 0.98679 | 0.98713 | 0.98745 | 0.98778 | 0.98809 | 0.98840 | 0.98870 | 0.98899 |
| **2.3** | 0.98928 | 0.98956 | 0.98983 | 0.99010 | 0.99036 | 0.99061 | 0.99086 | 0.99111 | 0.99134 | 0.99158 |
| **2.4** | 0.99180 | 0.99202 | 0.99224 | 0.99245 | 0.99266 | 0.99286 | 0.99305 | 0.99324 | 0.99343 | 0.99361 |
| **2.5** | 0.99379 | 0.99396 | 0.99413 | 0.99430 | 0.99446 | 0.99461 | 0.99477 | 0.99492 | 0.99506 | 0.99520 |
| **2.6** | 0.99534 | 0.99547 | 0.99560 | 0.99573 | 0.99585 | 0.99598 | 0.99609 | 0.99621 | 0.99632 | 0.99643 |
| **2.7** | 0.99653 | 0.99664 | 0.99674 | 0.99683 | 0.99693 | 0.99702 | 0.99711 | 0.99720 | 0.99728 | 0.99736 |
| **2.8** | 0.99744 | 0.99752 | 0.99760 | 0.99767 | 0.99774 | 0.99781 | 0.99788 | 0.99795 | 0.99801 | 0.99807 |
| **2.9** | 0.99813 | 0.99819 | 0.99825 | 0.99831 | 0.99836 | 0.99841 | 0.99846 | 0.99851 | 0.99856 | 0.99861 |
| **3.0** | 0.99865 | 0.99869 | 0.99874 | 0.99878 | 0.99882 | 0.99886 | 0.99889 | 0.99893 | 0.99896 | 0.99900 |
| **3.1** | 0.99903 | 0.99906 | 0.99910 | 0.99913 | 0.99916 | 0.99918 | 0.99921 | 0.99924 | 0.99926 | 0.99929 |
| **3.2** | 0.99931 | 0.99934 | 0.99936 | 0.99938 | 0.99940 | 0.99942 | 0.99944 | 0.99946 | 0.99948 | 0.99950 |
| **3.3** | 0.99952 | 0.99953 | 0.99955 | 0.99957 | 0.99958 | 0.99960 | 0.99961 | 0.99962 | 0.99964 | 0.99965 |
| **3.4** | 0.99966 | 0.99968 | 0.99969 | 0.99970 | 0.99971 | 0.99972 | 0.99973 | 0.99974 | 0.99975 | 0.99976 |
| **3.5** | 0.99977 | 0.99978 | 0.99978 | 0.99979 | 0.99980 | 0.99981 | 0.99981 | 0.99982 | 0.99983 | 0.99983 |

**TABLE 3**

Control Chart Factors

| Number of Observations In Subgroup | $\overline{X}$ Chart A2 | R Chart $D_3$ | R Chart $D_4$ | Sigma Chart $A_3$ | Sigma Chart $B_3$ | Sigma Chart $B_4$ | Individuals Chart $E_2$ | Median Chart $A_6$ |
|---|---|---|---|---|---|---|---|---|
| 2  | 1.880 | 0     | 3.267 | 2.659 | 0     | 3.267 | 2.660 | 2.230 |
| 3  | 1.023 | 0     | 2.575 | 1.954 | 0     | 2.568 | 1.772 | 1.187 |
| 4  | 0.729 | 0     | 2.282 | 1.628 | 0     | 2.266 | 1.457 | 0.829 |
| 5  | 0.577 | 0     | 2.115 | 1.427 | 0     | 2.089 | 1.290 | 0.691 |
| 6  | 0.483 | 0     | 2.004 | 1.287 | 0.030 | 1.970 | 1.184 | 0.561 |
| 7  | 0.419 | 0.076 | 1.924 | 1.182 | 0.118 | 1.882 | 1.109 | 0.509 |
| 8  | 0.373 | 0.136 | 1.864 | 1.099 | 0.185 | 1.815 | 1.054 | 0.441 |
| 9  | 0.337 | 0.184 | 1.816 | 1.032 | 0.239 | 1.761 | 1.010 | 0.412 |
| 10 | 0.308 | 0.223 | 1.777 | 0.975 | 0.284 | 1.716 | 0.975 | 0.369 |
| 11 | 0.285 | 0.256 | 1.744 | 0.927 | 0.321 | 1.679 | 0.946 | 0.350 |
| 12 | 0.266 | 0.284 | 1.716 | 0.886 | 0.354 | 1.646 | 0.921 | |
| 13 | 0.249 | 0.308 | 1.692 | 0.850 | 0.382 | 1.618 | 0.899 | |
| 14 | 0.235 | 0.329 | 1.671 | 0.817 | 0.406 | 1.594 | 0.881 | |
| 15 | 0.223 | 0.348 | 1.652 | 0.789 | 0.428 | 1.572 | 0.864 | |
| 16 | 0.212 | 0.364 | 1.636 | 0.763 | 0.448 | 1.552 | 0.849 | |
| 17 | 0.203 | 0.379 | 1.621 | 0.739 | 0.466 | 1.534 | 0.836 | |
| 18 | 0.194 | 0.392 | 1.608 | 0.718 | 0.482 | 1.518 | 0.824 | |
| 19 | 0.187 | 0.404 | 1.596 | 0.698 | 0.497 | 1.503 | 0.813 | |
| 20 | 0.180 | 0.144 | 1.586 | 0.680 | 0.510 | 1.490 | 0.803 | |
| 21 | 0.173 | 0.425 | 1.575 | 0.663 | 0.523 | 1.477 | 0.794 | |
| 22 | 0.167 | 0.434 | 1.566 | 0.647 | 0.534 | 1.466 | 0.785 | |
| 23 | 0.162 | 0.443 | 1.557 | 0.633 | 0.545 | 1.455 | 0.778 | |
| 24 | 0.157 | 0.452 | 1.548 | 0.619 | 0.555 | 1.445 | 0.770 | |
| 25 | 0.153 | 0.459 | 1.541 | 0.606 | 0.565 | 1.435 | 0.763 | |

# Glossary

**Accept:** To decide in a sampling process that an item, batch, or lot of a product or service satisfies the specified requirements.

**Acceptance Sampling:** A quality control procedure in which accept or reject decisions are made based on the proportion of defective items in a random sample.

**Accuracy:** The degree of closeness to the stated requirement.

**Actual Flowchart:** A graphical representation of how a process works in reality.

**Addition Principle of Probability:** If one group contains "x" items, another group contains "y" items, and "z" represents the items that are in both groups, then the number of items that are in both two groups is x + y − z. Likewise, when dealing with probabilities, the equation becomes (where E1 is the first event and E2 is the second event):

$$Pr\{E1 \text{ or } E2\} = Pr\{E1\} + Pr\{E2\} - Pr\{E1 \text{ and } E2\}$$

**American National Standards Institute (ANSI):** The coordinator for establishing voluntary standards.

**American Society for Quality (ASQ):** A professional organization that promotes quality related information and techniques for the academia, government, and industrial sectors.

**Analysis:** An investigation to find the problem, root cause, and others by applying a systematic approach.

**Area of Opportunity Control Charts:** Control charts that deal with the number of times a particular event occurs in a given area of opportunity.

**Arithmetic Average:** A single number representing a group of numbers. An average is calculated by adding all the individual items of a sample or population and then dividing this number by the total number of items. Also called average or mean.

**Assignable Cause Variation:** Product or process variation that exists because of special circumstances that are not inherent to the product or process. Also called special cause variation.

**Attribute Control Chart:** A statistical process control chart of attributes data of a process.

**Attribute Data:** Data that have to do with the number of times a particular event occurs; for example, the number of defective units in a lot or the number of defects in a particular unit.

**Audit Trail:** The flow of events and process improvement changes that are documented such that the following individual or group can clearly understand what has happened to the process.

**Average:** *See* Arithmetic Mean.

**Average and Range Chart:** A variables statistical process control chart that depicts the central tendency and variability of a process over time.

**Average and Sigma Chart:** A variables statistical process control chart that depicts the central tendency and standard deviation of a process over time. The average and sigma chart is used when the subgroup size is ten or more.

**Baldrige, Malcolm:** The late Secretary of Commerce who was an advocate of quality.

**Baseline Measure:** With respect to processes, the beginning statistics, inclusive of central tendency and variability measures, that depict what the process is doing.

**Bell-Shaped Curve:** The shape of a histogram that exhibits normalcy.

**Benchmarking:** A process improvement technique that consists of comparing organizational, performance, financial, and other indicators to the best-in-class organization.

**Best Flowchart:** A graphical representation of the optimum flow of a process.

**Bias:** In surveys, an error that is introduced when the replies of respondents are influenced in a one-sided manner by the way the survey questions are worded.

**Bimodality:** A situation that arises when two modes occur in a given distribution.

**Binomial Count Control Charts:** Control charts that deal with the number or fraction of items in a subgroup that meet a particular event.

**Brainstorming:** A creative thinking technique to generate free and open ideas within a group.

**Business Operations:** Similar to processes. In many texts, defined as the transformation of resources (inclusive of manpower, materials, methods, and machinery) into products and services.

**Calibrate:** To correct a measuring instrument to an accepted standard.

**Capability:** The ability of a process to deliver a product or service within the specified requirements.

**Cause:** The root reason for a deviation between the actual results and the expected results.

**Cause-and-Effect Diagram:** A graphical representation of the effect and the causes that contribute to that effect. Also called the Ishikawa diagram and Fishbone diagram.

**c Charts:** An area of opportunity control chart that counts the number of nonconformities per unit.

**Centerline:** In statistical process control charting, the overall average operating level of the process.

**Central Tendency:** A mathematical concept describing where the data from a sample or population tend to center themselves. The measures of central tendency are the mean, median, and mode.

**Characteristic:** A measurable property of an item, sample of items, or population of items.

**Chart:** A graphical display depicting data.

**Checklist:** A method to ensure that important steps are taken.

**Common Cause Variation:** A process inherent source of variation.

**Compliance:** A condition that exists when a product or service meets the specified requirement or standard.

**Conditional Probability:** Frequently there exists a need to determine the probability of an event occurring given the fact that some other event has already occurred. This type of probability is called conditional probability and is expressed as Pr{A/B}, where A and B are different events. In words it is expressed as "the probability that event A will occur GIVEN the fact that event B has already occurred."

**Confidence Level:** The certainty level that the measured probability of some assertion is true.

**Conformance:** *See* Compliance.

**Consensus:** A state where everyone in the group supports a decision or course of action, even if some do not totally agree.

**Consistency:** A degree of uniformity with little or no variation.

**Consumer:** The end user of a product or service.

**Continuous Improvement:** The never-ending betterment of a product or service via increased process performance.

**Control Chart:** In statistical process control, a graphical chart of the variation exhibited by a product or process over time.

**Control Chart Factors:** Numeric values used in the calculation to determine the upper and lower control limits for various control charts.

**Control Limits:** In statistical process control, the boundaries of normal variation for a product/process. Mathematically speaking, the control limits are set at the mean + 3 standard deviations and the mean − 3 standard deviations.

**Corrective Action:** An action that is taken to eliminate the causes of nonconformity after the nonconformity has already occurred.

**Cost of Quality:** The expense of product/service failure. Usually categorized as preventive, appraisal, internal failure, and external failure. Also called quality costs.

**Cost Pareto:** A Pareto chart that displays the costs of non-quality as opposed to defect categories.

**Cp:** A process capability index that compares the specification width to the process width. Cp is strictly a variation index.

**Cpk:** A process capability index that considers both process centering and variability to the specification.

**Crosby, Philip:** The founder of Philip Crosby Associates, Inc. and Crosby Quality College in Orlando, Florida. In the 1980s he led and participated in many high-profile initiatives. Philip Crosby's philosophy focused around changing management's posture toward a higher standard of performance, namely "zero defects."

**Cumulative:** The successive summing of the parts.

**Customer:** The person receiving a value added product or service.

**Customer Acceptance:** The buyer act of formally accepting a product or service as meeting the specified requirement.

**Customer Requirements:** Performance standards associated with specific customer needs.

**Customer Requirements Checklist:** A list of common questions that serves as a basis for developing customer requirements.


**Data:** A term used to denote some representation or information.

**Data Source:** The origin of the collection of information.

**Defect:** A state or condition that is usually associated with the evaluation of product/service with respect to usability.

**Defective Unit:** Usually identified as a unit that is unusable due to one or more nonconformities.

**Degree of Confidence:** *See* Confidence Level.

**Delinquency:** The failure to meet a specified requirement.

**Deming, W. Edwards:** A renowned worldwide management consultant for nearly 40 years. He graduated in 1928 with a Ph.D. in physics from Yale University and worked with Walter Shewhart (designer of the control chart in 1924) at Bell Labs and the Hawthorne Plant. He is and has been a Japanese hero, although not all of American industry has been totally introduced to his philosophy.

**Design Array:** A predetermined array that represents the experimental settings for an experiment.

**Design of Experiments (DOE):** The technique of manipulating controllable factors in order to not only determine their effect on some desired outcome, but to find the optimum combination of factors that will yield the desired outcome. Same as experimental design.

**Detection:** A reactive strategy that attempts to identify and correct a faulty product/service after it has been produced.

**Deviation:** The difference between the actual results and expected results.

**Diagnosis:** The act of discovering the cause(s) of nonconformities.

**Dirty Data:** Collected data that do not represent what they are intended to represent.

**Discrimination:** The degree of fineness of the measurement.

**Dispersion:** A degree of variation.

**Distribution:** A graphical chart depicting the arrangement of data.

**Documentation:** The printed material depicting the process, what is happening in the process, and any changes that are made to the process.

**Effect:** The result of some action.

**Eighty/Twenty Rule:** The Pareto Principle where 80% of the variation or results are explained by 20% of the causes.

**Empowerment:** The state where employees have the authority to make decisions and take action without approval, and likewise possessing the accountability and responsibility for the decisions and their outcomes.

**End User:** The recipient of the final product or service. Also called consumer.

**Experiment:** A designed test or trial under specified settings.

**Experimental Design:** *See* Design of Experiments.

**External Customer:** The end user of a product or service.

**Factor:** A cause or setting that will influence the response or output.

**Failure:** The state that exists when an item does not meet or perform as intended.

**Feedback:** The return of information about a product, service, or process that is used to improve the current state.

**Final Inspection:** The act of checking, analyzing, and examining the finished product or service before delivery to the customer.

**Fishbone Diagram:** *See* Cause-and-Effect Diagram.

**Flowchart:** A graphical representation of the specific steps, or activities, of a process.

**Fourteen Points:** W. Edwards Deming's management philosophy to help companies increase their quality and productivity.

**Fractional Factorial Experiments:** Experiments that require a fraction of the total experiments associated with a full factorial experiment.

**Frequency:** In statistics, the number of times something occurs.

**Frequency Distribution:** A graph depicting the number of occurrences of a variable.

**Full Factorial Experiment:** An experimental design that considers all possible combinations of factors and levels.

**Glitch:** A temporary erroneous or nonconforming response.

**Goal:** A specific and measurable statement of the desired outcome.

**Go/no-go:** A condition where the output is categorized in one of two groups: conforming or nonconforming.

**Graph:** A pictorial representation of data.

**Hierarchy:** An arrangement, or ordering, into a progressive series.

**Histogram:** A graph, or frequency distribution, of a variable.

**Hypothesis:** A yet to be proven statement.

**Implementation:** The movement of a design into reality.

**Implied Warranty:** An assurance by the producer to the buyer that the product or service is reasonably fit for its intended use.

**In-Control:** A statistical state denoting the operation of a process within the boundaries of the mean ± 3 standard deviations.

**Individuals and Range Chart:** A variables statistical process control chart that depicts the individual data points and moving range. This chart is used when there are large amounts of time between data collection, or data collection requires the destructive testing of a product, which may prove economically unfeasible.

**Information:** The conversion of data into useful forms, or concepts, about a product, service, process, or population.

**Inherent Capability:** The range of variation of the process that is characterized as the area between the mean − 3 standard deviations and the mean + 3 standard deviations.

**Input:** As used in productivity, the resources needed to produce a product or service.

**Inspection:** The act of measuring, checking, analyzing, examining, and testing characteristics of an item, product, or process and comparing that result to specified requirements to determine a degree of conformity.

**Inspection Rectification:** The act of replacing a faulty unit (when discovered through inspection) with a good unit to replace it.

**Intermittent:** Not occurring regularly.

**Internal Customer:** The next person in the organization who receives your product/service.

**International Organization for Standardization (ISO):** A society that has been developing standards for international companies since approximately 1947. ISO consists of national institutes from countries throughout the world.

**Ishikawa, Kaoru:** Creator of the cause-and-effect diagram in 1943. Dr. Kaoru Ishikawa's quality technique gained widespread usage throughout Japan and the world.

**Ishikawa Diagram:** *See* Cause-and-Effect Diagram.

**Juran, Joseph M.:** A quality guru who was directly responsible for several high-profile successful programs in the 1980s, and his three books (*Quality Control Handbook, Quality Planning and Analysis,* and *Management of Quality Control*) have collectively been translated into twelve different languages. He (like Dr. Deming) worked with Walter Shewhart in the Hawthorne Plant in 1924.

**Learning Curve:** A line graph usually depicting the relationship between consecutive number of units produced (X-axis) and the time required to produce them (Y-axis).

**Likelihood:** The probability that an event will occur.

**Limit:** A boundary or condition (for example, upper and lower control limit, and upper and lower tolerance limit).

**Logic:** Reasoning.

**Lower Control Limit (LCL):** In statistical process control, the lower boundary for the area of statistical control. Mathematically defined as the mean − 3 standard deviations.

**Lower Specification Limit (LSL):** Same as lower tolerance limit (LTL).

**Lower Tolerance Level (LTL):** The lower boundary of acceptance for a stated requirement. Same as lower specification limit (LSL).

**Malcolm Baldrige Award:** An award that promotes the awareness and importance of quality in an effort to increase U.S. companies' competitive position on a worldwide basis.

**Malcolm Baldrige Award Criteria:** The categories and elements of the Malcolm Baldrige Award. A company is awarded points based on how well they satisfy the criteria.

**Management by Crises:** A reactive leadership style that supports dealing with problems after they become critical.

**Margin of Error:** The ± range of the results of data collected from a sample.

**Mean:** Same as arithmetic average.

**Measurement:** To proportion by some measured lot.

**Measure of Central Tendency:** A statistical measure of where the data tend to center themselves. The measures of central tendency discussed in this text are the mean, median, and mode.

**Median:** The value of the middle item when all of the items in the group are arranged in ascending or descending order.

**Median and Range Chart:** A variables statistical process control chart that depicts the median and range of the process. This chart is much easier to build in terms of mathematics and is recommended as an introduction to control charting, with progression to the other variables control charts.

**Metrology:** The science of measurement.

**Mode:** The most frequently occurring item in a group of data.

**Ms of a Process:** The inputs or resources of a process, such as manpower, methods, machinery, monies, materials, and message media.

**Mutually Exclusive:** A probability condition where two or more events may occur, but the occurrence of one precludes the occurrence of the other(s).

**Noise:** In design of experiments, the effect on the output characteristic caused by uncontrollable factors.

**Nominal:** An intended amount or condition.

**Nonconforming Unit:** A product or service that contains one or more nonconformities.

**Nonconformity:** A state or condition that occurs when the specified requirement is not met.

**Non-Value Added:** An activity that adds nothing to the product or service.

**Normal Curve:** A symmetrical bell-shaped histogram of the frequency of some variable. Also called the bell-shaped curve.

**np Chart:** A binomial count control chart that is used to count the number of nonconforming items in a uniform sample size.

**Obsolescence:** In the process of going out of date.

**One Factor at a Time Experiment:** An experiment in which a series of factors are studied, but only one factor at a time is adjusted.

**Optimal:** The maximum operating level in terms of specified requirements.

**Outcome:** The results of a trial, exercise, or experiment.

**Outlier:** An extreme observation that is definitely considered as being questionable.

**Out-of-Control:** In statistical process control, when the process is not in a state of control (mean ± 3 standard deviations).

**Paradigm:** A common set of beliefs or assumptions held by a particular group or team.

**Pareto Diagram:** A bar chart with the defect categories shown in descending order (from highest category to lowest).

**Pareto Law:** Same as Pareto Principle.

**Pareto Principle:** The concept that 80% of the variation is caused by 20% of the sources.

**Pareto Rule:** Same as Pareto Principle.

**Pareto, Vilfredo:** An economist who discovered that 80% of the wealth lies in 20% of the populations' hands.

**Pattern:** In statistical process control, a repeating series of events, shape, etc.

**p Chart (or Percent chart):** A binomial count control chart that is used for proportions, or proportion nonconforming.

**Performance Specification:** A statement of the requirements.

**Plan/Do/Check/Act Cycle or Plan/Do/Show/Act Cycle:** Sometimes called the Deming Cycle or Shewhart Cycle. The four-step process improvement technique that can be used at all levels and at all stages of the process as guidelines for improvement.

**Precision:** The dispersion, or variability, of measurement.

**Prevention:** A proactive strategy that attempts to identify and correct a faulty product or service before it ever has been produced (for example, during the design phase or development phase).

**Preventive Action:** An action that is taken to eliminate the causes of nonconformity before the nonconformity has occurred.

**Probability:** The likelihood that an event will occur.

**Probability of Acceptance:** The likelihood that an item, batch, or lot will be accepted when measured against the stated requirement.

**Process:** A set of usually sequential value-added tasks that use organizational resources to produce a product or service.

**Process Analysis:** The systematic study of the process and what is happening in the process.

**Process Capability:** The ability of a process to meet the customer requirement.

**Process Capability Indices:** A technique for specifying quality by statistical measures. They are another way to quantify the degree to which a process is, or is not, meeting the customer requirement. The two process capability indices studied in this text are Cp and Cpk.

**Process Checklist:** A list of questions used to assist in the understanding of a process. The process checklist should be used before any process improvement initiatives are implemented.

**Process Control:** Same as statistical process control. The act of analyzing the variation associated with a process to determine if a state of statistical control exists.

**Productivity:** A ratio measuring the process output to the process inputs.

**Qualitative:** The inductive analytical approach as relating to nonmeasurable, subjective values.

**Quality:** Giving the customer the right thing right the first time.

**Quality Assurance:** Every planned task and action necessary that demonstrate that the product or service satisfies the given customer requirements. In other words, everything is done to ensure that quality control is what it should be.

**Quality Control:** All operational techniques that are necessary to satisfy all quality requirements. Inclusive in quality control is process monitoring and the elimination of root causes of unsatisfactory product/service quality performance. Quality Control usually refers to the activities that occur on the shop floor to maintain the quality level of a product.

**Quality Trilogy:** The Dr. Juran proposed three-pronged approach (quality planning, quality control, and quality improvement) for the management of quality.

**Quantitative:** The deductive analytical approach as relating to measurable values.

**Questionnaire:** A form for soliciting and recording data.

**Random:** Existing or occurring by chance.

**Random Cause Variation:** Same as common cause.

**Random Sample:** A sample where all items in the population have an equal probability of being selected.

**Range:** In statistics, a measure of variability calculated by taking the difference between the highest value and the lowest value.

**Range Chart:** In statistical process control, a graph indicating the changes in variation of a product/service.

**Ratio:** A relationship of two numbers, usually the result of dividing one number into another.

**Raw Data:** Original information about something in its unsorted or untouched form.

**R-chart:** Same as range chart.

**Receiving Inspection:** The analyzing, checking, examining, and testing of incoming materials from various suppliers.

**Rectification:** Replacement of a bad unit with a good one, as in Inspection Rectification.

**Reliability:** In a product or process, the item functions as it is intended to function without failure for a specified interval of time under specified conditions.

**Representative Sample:** In sampling, a group of items that depict what is going on in the population.

**Requirement:** A condition, specification, or capability needed by a customer; usually termed customer requirement.

**Risk:** The probability of making an incorrect decision.

**Robust Design:** A design that is immune, or insensitive, to noise factors.

**Root Cause:** The base reason for the failure, defect, or nonconformity.

**Run Chart:** A graphical display of data in the order that they occur.

**Sample:** In statistics, a representative subgroup of a population.

**Sampling:** The act of randomly selecting items from a population.

**Scatter Diagram:** An X-Y graphical plot used to analyze the relationship between two variables, with one variable represented on the X-axis and one variable represented on the Y-axis.

**Sequential:** Ordered in a timed manner.

**Six-Sigma:** A technique which, when correctly implemented, yields 3.4 parts per million defects (99.9996% of the process output lies between the upper and lower specification limits).

**Skew:** The lack of symmetry.

**Special Cause Variation:** Variation that occurs because of special circumstances, or not inherent to the process. The area of special cause variation is above the upper and below the lower control limit boundaries.

**Specification:** Specific and measurable attributes that convey the customer requirements.

**Stable Process:** The condition that exists when a process is plotted on a statistical process control chart and shows only random patterns with no data points outside of the upper and lower control limits (mean ± 3 standard deviations).

**Standard Deviation:** A statistical measure of variability.

**Statistical Analysis:** The collecting and analyzing of data using statistical techniques as their basis.

**Statistical Control:** In statistical process control, a state that occurs when all data points lie between the upper and lower control limits (mean ± 3 standard deviations). Same as stable process.

**Statistical Process Control:** The application of statistical techniques and charting that provides insight into the control of variation.

**Statistical Quality Control:** Applications of statistical techniques inclusive of statistical process control, sampling plans, diagnostics tools, etc.

**Statistical Sampling:** Sampling using statistical techniques with a given confidence level and margin of error.

**Survey:** A questionnaire given to respondents to determine their opinions regarding the quality of the product or service.

**System:** A group of interconnected elements that can operate in a self-sufficient mode of operation to accomplish some function.

**Task:** A specific definable activity to perform a portion of a process.

**Taguchi, Genichi:** A Japanese quality expert involved in quality improvement. He introduced several statistical techniques that have been proven to be of extreme value, and emphasized with his techniques the building of quality into the design, products, and processes.

**Theoretical Flowchart:** A graphical representation that shows the way a process should operate according to some prescribed policy, procedure, or operations manual.

**Tolerance:** The acceptable range of variation permitted by the specification.

**Tolerance Stack-Up:** A condition that occurs when the tolerances of the individual parts combine together (when assembled) to result in a component tolerance that may be out of tolerance for that assembly.

**Total Quality Management (TQM):** An integrated system for achieving customer satisfaction that involves all members of the organization and uses quantitative techniques to continually improve the processes.

**Trial And Error Experiment:** A type of experiment where one factor is manipulated without regard to the other factors.

**Tweaked:** The attempt to improve a situation without quantitative information to support the improvement action. Rather, reliance on intuition is the basis for process adjustments.

**u Charts:** An area of opportunity control chart that is similar to the c-Chart with the exception that the subgroup size can vary.

**Upper Specification Limit (USL):** The upper boundary of acceptance for a stated requirement. Same as upper specification limit (USL).

**Upper Tolerance Limit (UTL):** Same as upper specification limit (USL).

**Value-Added:** An activity that adds a benefit or utility to the product or service.

**Variability:** Dispersion or lack of consistency around a target, objective, goal, etc.

**Variables Control Chart:** A graphical representation in statistical process control of the variables data of a process.

**Variables Data:** Data that are associated with measuring items.

**Variation:** Same as variability.

**Warranty:** An agreement provided by the seller that the product or service will perform as intended for a specified interval of time for a stated period of time.

**World Class:** The "best of the best" when considering quality.

**X-bar and R-Chart:** Same as average and range chart.

**X-bar and s-Chart:** Same as average and sigma chart.

**Z-test:** In statistics, the term given to the test to calculate the area under a curve to the left of some point of interest.

**Zero Defects:** A long-range concept associated with Philip Crosby, intending to mean the never-ending quest for improvement.

# Index